Math Everywhere
ONTARIO EDITION
5

SENIOR MATH EVERYWHERE AUTHORS
Peter Rasokas, Barry Scully, Jan Scully, Bryan Szumlas

MATH EVERYWHERE K–8 AUTHOR TEAM

Brendene Barkley	Betty McKendry
Luisa Busato	Linda Miller
Geoff Cainen	Emma Mills Mumford
Tara Cook	Carla Pieterson
Garey Edgar	Pamela Quigg
Donna Green	Maureen Rousseau
Liz Holder	Robert Stoddart
James King	Lori Wiens
Barbara E. Worth	

CONSULTANTS

ASSESSMENT CONSULTANTS **Kelly Lantink, Kevin Akins**
EDITORIAL CONSULTANT **Mary Jean Tyczynski**
MATHEMATICS DEVELOPMENT CONSULTANTS **Mary Ellen Diamond, Kathleen Nolan**
MATHEMATICS LITERACY CONSULTANT **Cathy Marks Krpan**
TECHNOLOGY CONSULTANTS **Marilyn Legault, Doug McKnight**
ADVISOR ON ABORIGINAL PERSPECTIVES **Ken Ealey, Sonja Willier**

Elementary Education Advisory Board
CURRICULUM ADVISOR **Les Asselstine**
LITERACY ADVISOR **David Booth**
STAFF DEVELOPMENT ADVISOR **Rod Peturson**

Harcourt Canada

Orlando   Austin   New York   San Diego   Toronto   London

Copyright © Harcourt Canada Ltd.

All rights reserved. No part of this publication may be reproduced or transmitted in any form or by any means, electronic or mechanical, including photocopy, recording, or any information storage and retrieval system, without permission in writing from the publisher. Reproducing passages from this book without such written permission is an infringement of copyright law.

Requests for permission to photocopy any part of this work should be sent in writing to: Schools Licensing Officer, access©, 1 Yonge Street, Suite 1900, Toronto, ON, M5E 1E5. Fax: (416) 868-1621. All other inquiries should be directed to the publisher.

Every reasonable effort has been made to acquire permission for copyright material used in this text, and to acknowledge such indebtedness accurately. Any errors or omissions called to the publisher's attention will be corrected in future printings.

National Library of Canada Cataloguing in Publication Data
    Math everywhere 5/senior author and grades 4–6 coordinator
    Peter Rasokas; authors: Linda Miller ... [et al.].

ISBN 0-7747-1539-1

    1. Mathematics—Textbooks. I. Rasokas, Peter, 1949– II. Miller, Linda, 1951–

QA107.2.M3845 2002   510   C2002–903722-0

Editorial Project Manager: Ian Nussbaum
Developmental Editors: Sasha Patton, Brett Savory, Camille Isaacs, Todd Mercer,
    Erynn Prousky, Elizabeth Salomons
Production Coordinator: Cheri Westra
Production Assistant: Agnieszka Mlynarz
Permissions Editor and Photo Research: Karen Becker
Art Direction and Design: Sonya V. Thursby/Opus House Incorporated
Composition: Susan Purtell
Printing and Binding: Transcontinental Printing Inc.
Cover Image: © Digital Vision/Getty Images

∞ Printed in Canada on acid-free paper.
1 2 3 4 5   07 06 05 04 03

Welcome to *Math Everywhere!*

Did you know math is all around you? Math is in your classroom, it's in your home, and it's in your city or town.

Math has also been used throughout history and can be used to plan the future.

At the beginning of this year, in **Start-Up Math**, you will review what you already know about math and prepare for the school year.

In **Unit 1, The Mathematics of Weather**, you will learn about numbers and patterns as you explore weather.

In **Unit 2, Math From the Past**, you will learn about shapes and measurement, and about numbers and patterns as you visit ancient civilizations.

In **Unit 3, Math in Energy**, you will learn about numbers, data, patterns, and measurement as they relate to energy and energy conservation.

At the end of the year, in **Celebrating Math**, you will use what you have learned about math this year to plan a bake sale.

At the end of each chapter, you will review what you have learned and show your understanding of math by doing a wrap-up activity.

You will play games, solve riddles, and solve problems while learning all about math!

You will see yellow shapes in some parts of the book:
- tells you that you will be solving a problem.
- lets you know that your answers will show what you understand about math.
- shows questions that will let you apply what you know about math.
- tells you that you will communicate what you know about math.

We hope you enjoy using this book and that you have fun learning that math **is** everywhere.

Contents

Start-Up Math

Start-Up Math introduces you to *Math Everywhere* and lets you show what you already know about math.

Lesson 1: What Is Math	2
Lesson 2: Number Sentences	4
Lesson 3: Making Geometric Figures	6
Lesson 4: Measuring With String and Cubes	9

Problems to Solve   11

Problem 1: Math at the Movies	11
Problem 2: Family Reunion Plan	13
Problem 3: Counting Animals	15
Problem 4: Estimating and Checking on a Nature Walk	16
Problem 5: Backwards Bricklaying	18

Unit 1

The Mathematics of Weather

Chapter 1
Weather Numbers   21

This chapter focuses on *Number Sense and Numeration*.

Lesson 1: Exploring Place Value	22
Lesson 2: Exploring Whole Numbers	26
Lesson 3: Exploring Decimal Numbers	29
Show What You Know: Lessons 1 to 3, Whole Numbers and Decimal Numbers	**33**
Lesson 4: Exploring Number Patterns	34
Lesson 5: Adding and Subtracting Whole Numbers	37
Lesson 6: Multiplying and Dividing Whole Numbers	42
Show What You Know: Lessons 4 to 6, Operations With Whole Numbers	**45**
Lesson 7: Adding and Subtracting Decimal Numbers	46
Lesson 8: Multiplying Decimal Numbers	50

Lesson 9: Dividing Decimal Numbers	53
Lesson 10: Mental Math Strategies	56
Show What You Know: Lessons 7 to 10, Mental Math and Operations With Decimals	**59**
Lesson 11: Solving Two-Step Number Problems	60
Chapter Review	64
Chapter Wrap-Up	66

Chapter 2
Weather Data   68

This chapter focuses on *Measurement* and *Data Management and Probability*.

Lesson 1: Converting Metric Measurements	69
Lesson 2: Reading Analog Clocks	73
Lesson 3: Using SI Notation	76
Show What You Know: Lessons 1 to 3, Measurement and Time	**79**
Lesson 4: Exploring Money	80
Lesson 5: Data Collection and Analysis	85
Lesson 6: Line Graphs	90
Show What You Know: Lessons 4 to 6, Money and Data Management	**93**
Lesson 7: Bar Graphs	94
Lesson 8: Pictographs	98
Lesson 9: Calculating Mean	102
Show What You Know: Lessons 7 to 9, Data Management	**106**
Lesson 10: Predicting From Patterns	107
Chapter Review	111
Chapter Wrap-Up	114

Chapter 3
Weather Patterns   116

This chapter focuses on *Patterning and Algebra* and *Number Sense and Numeration*. This chapter also touches upon *Measurement* and *Geometry and Spatial Sense*.

Lesson 1: Extending and Creating Patterns	117
Lesson 2: Number Position and Value	121
Lesson 3: Number Pattern Relationships	125
Show What You Know: Lessons 1 to 3, Patterns	**129**
Lesson 4: Multiplication and Division Patterns	130
Lesson 5: Patterns With Variables	134
Lesson 6: Exploring Pattern Rules	138
Show What You Know: Lessons 4 to 6, Patterns	**142**
Lesson 7: Finding Missing Factors	143
Lesson 8: Changes in Factors	147
Show What You Know: Lessons 7 and 8, Factors	**150**
Lesson 9: Three-Dimensional Patterns	151
Chapter Review	155
Chapter Wrap-Up	158

Problems to Solve	**160**
Problem 6: Target Data	160
Problem 7: Building Measurements	162
Problem 8: Money Patterns	164
Problem 9: Fish Ladder Patterns	166

Unit 2

Math From the Past

Chapter 4
Patterns From the Past — 169

This chapter focuses on *Geometry and Spatial Sense* and *Number Sense and Numeration*. This chapter also touches upon *Measurement* and *Patterning and Algebra*.

Lesson 1: Identifying and Describing Polygons	170
Lesson 2: Classifying Triangles by Side Lengths	173
Lesson 3: Classifying Triangles by Angles	178
Show What You Know: Lessons 1 to 3, Identifying Polygons, Classifying Triangles	**182**
Lesson 4: Identifying Congruent Shapes and Angles	183

Lesson 5: Identifying and Comparing Equivalent Fractions	188
Lesson 6: Using Multiplication to Create Equivalent Fractions	192
Lesson 7: Exploring Fractions and Decimals to Tenths	196
Lesson 8: Exploring Fractions and Decimals to Hundredths	200
Show What You Know: Lessons 4 to 8, Identifying Congruent Shapes and Angles, Exploring Fractions and Decimals	**204**
Lesson 9: Repeating and Expanding Patterns	205
Lesson 10: Identifying, Extending, and Creating Numbers Patterns	208
Chapter Review	212
Chapter Wrap-Up	214

Chapter 5
Math in Ancient Cities — 217

This chapter focuses on *Measurement* and *Geometry and Spatial Sense*. This chapter also touches upon *Patterning and Algebra* and *Data Management and Probability*.

Lesson 1: Measurement—Kilometres	218
Lesson 2: Estimating and Measuring Dimensions of Objects	221
Lesson 3: Relationships Between Metres and Centimetres	226
Lesson 4: More Relationships Between Units of Measurement	230
Show What You Know: Lessons 1 to 4, Measurement	**234**
Lesson 5: Calculating Perimeter	235
Lesson 6: Perimeter—Irregular Shapes	239
Show What You Know: Lessons 5 and 6, Perimeter	**241**
Lesson 7: Finding the Area of Irregular Shapes	242
Lesson 8: Measurement Relationships and Formulas	247

Lesson 9: Changing the Dimensions of a Rectangle	250
Lesson 10: Rectangles, Squares, and Diagonals	253
Show What You Know: Lessons 7 to 10, Measurement	**256**
Chapter Review	257
Chapter Wrap-Up	260

Chapter 6
Math in Monuments and Marvels — 262

This chapter focuses on *Geometry and Spatial Sense*. This chapter also touches upon *Number Sense and Numeration, Measurement, Patterning and Algebra,* and *Data Management and Probability*.

Lesson 1: Estimating and Measuring Heights	263
Lesson 2: Comparing Pyramids and Prisms	267
Lesson 3: Pyramids, Prisms, and Nets	271
Lesson 4: Exploring the Faces of Pyramids and Prisms	275
Lesson 5: Creating and Solving Problems About the Great Pyramid	279
Lesson 6: Constructing Solids Made From Cubes	282
Show What You Know: Lessons 1 to 6, Measuring and Pyramids, Prisms, and Solid Figures	**284**
Lesson 7: The Tangram Puzzle	285
Lesson 8: Exploring Tangrams Further	288
Lesson 9: Probability	291
Lesson 10: Exploring Probability Further	295
Show What You Know: Lessons 9 and 10, Probability	**298**
Chapter Review	299
Chapter Wrap-Up	302

Problems to Solve — 303

Problem 10: Making Equations	303
Problem 11: Using Logic	305
Problem 12: Estimating Lengths of Bathroom Tissue	306
Problem 13: A Division Problem	308
Problem 14: Working With Whole Numbers	309

Unit 3

Math in Energy

Chapter 7
Electric Math   311

The chapter focuses on *Data Management and Probability* and *Number Sense and Numeration*.

Lesson 1: Analyzing Energy Data	312
Lesson 2: Working with Data and Formulas	316
Lesson 3: Examining and Organizing Energy Data	319

Show What You Know: Lessons 1 to 3, Analyzing Data   322

Lesson 4: Calculating Energy Cost	323
Lesson 5: Analyzing an Energy Graph	327
Lesson 6: Working With Money	330

Show What You Know: Lessons 4 to 6, Data Management and Money   332

Lesson 7: Multiplying by 10, 100, and 1000	333
Lesson 8: Looking at Energy Tables	338
Lesson 9: Comparing Data	341
Lesson 10: Calculating Costs	344

Show What You Know: Lessons 7 to 10, Multiplication and Calculating Costs   347

Lesson 11: Finding Perimeter and Area	348
Chapter Review	351
Chapter Wrap-Up	353

Chapter 8
Seedling Sums   355

This chapter focuses on *Number Sense and Numeration*, *Patterning and Algebra*, and *Data Management and Probability*.

Lesson 1: Calculating Tree Costs	356
Lesson 2: Fractions of Whole Numbers	358

Lesson 3: Calculating Your Earnings	362
Lesson 4: Transformations, Fractions, and Patterns	364
Show What You Know: Lessons 1 to 4, Cost, Fractions, and Geometry	**367**
Lesson 5: Water Volume and Capacity	368
Lesson 6: Collecting and Organizing Paper Data	370
Lesson 7: Data With Fractions and Decimals	372
Lesson 8: Surveys and Graphs	375
Show What You Know: Lessons 5 to 8, Measurement, Numbers, and Data	**378**
Lesson 9: Patterns and Tessellations	379
Lesson 10: Analyzing and Graphing More Data	382
Lesson 11: Examining Data From a Table	385
Chapter Review	388
Chapter Wrap-Up	390

Chapter 9
Energy by the Numbers — 392

This chapter focuses on *Data Management and Probability* and *Number Sense and Numeration*.

Lesson 1: Predicting Probability	393
Lesson 2: Using Multiplication to Analyze Data	396
Lesson 3: Working With Water Data	399
Lesson 4: Comparing Data Using Multiplication	402
Show What You Know: Lessons 1 to 4, Probability and Multiplying Data	**404**
Lesson 5: Problem Solving With Patterning	405
Lesson 6: Finding Patterns	408
Lesson 7: Probability Problems	411
Show What You Know: Lessons 5 to 7, Patterns, Money, and Probability	**413**
Lesson 8: Ranking and Graphing Data	414
Lesson 9: Analyzing and Comparing Data	417
Lesson 10: Transportation Survey and Probability	421
Chapter Review	424
Chapter Wrap-Up	427

Problems to Solve — 428
Problem 15: The Human Tower — 428
Problem 16: Capacity and Volume — 430
Problem 17: Go With the Flow — 432
Problem 18: How Old is Grandma? — 433

Celebrating Math

Celebrating Math lets you know what you have learned about math while celebrating the end of the year.

Lesson 1: Working With Division — 435
Lesson 2: Perimeter, Area, and Volume — 437
Lesson 3: Fractions and Decimals — 439
Lesson 4: Exploring Large Numbers — 441
Lesson 5: Counting Money — 443
Wrap-Up: Planning a Class Bake Sale — 445

Glossary — 447
Problem-Solving Strategies — 457
Acknowledgements — 459

Start-Up Math

Welcome to *Math Everywhere*. You are starting a new year of learning about math. Starting something new can sometimes be challenging, but it can also be exciting! During the next few days you will use what you already know to complete several mathematics activities. These activities help you check your understanding of math concepts and prepare you for the work that you do for the rest of the year.

To get you thinking about math, brainstorm as many possibilities as you can to answer the following question:

If 100 is the answer, what is the question?

NUMBER SENSE AND NUMERATION

Lesson 1

What Is Math?

Get Started

The Get Started section gives you a few quick activities to prepare you for the math lesson.

1. What does "math is everywhere" mean?
2. Look around your classroom and record examples of math. For example, there might be 40 tiles on the ceiling. The ceiling tiles make a pattern.

Build Your Understanding

Create a Math Collage

The Build Your Understanding title gives you a clue about what you will be learning or doing in the lesson.

You Will Need
- poster paper
- glue
- scissors
- old newspapers and magazines
- felt pens

Work with a partner.
1. Look through newspapers and magazines for examples of math.
2. Cut out and glue your findings to your poster paper.
3. Use the felt pens to highlight where the math is in your examples. On the back of your collage, explain why you cut out what you did.

2 Start-Up Math

What Did You Learn?

The What Did You Learn? sections are designed to help you think about what you have learned in the lesson.

1. Share, compare, and discuss your math collage with another pair of students. Then post it on a bulletin board. Don't forget to put your name on your poster.
2. Think about your math collage. Then, in your math journal, write about your math strengths and areas for improvement.

Practice

The Practice sections include many different types of activities that you can do to review what you have learned, or to challenge yourself further.

1. Write math sentences for which the solution is your favourite number. Try to use all four operations (addition, subtraction, multiplication, and division).
2. Write a paragraph about the ways you learn best in math.

NUMBER SENSE AND NUMERATION
GEOMETRY AND SPATIAL SENSE

Lesson 2

Number Sentences

Get Started

Make math sentences for each of the following answers. For example, if the number 10 is shown, one math sentence could be 5 x 2 = 10. Be sure to use all four operations.

1. 15
2. 20
3. 48
4. 99
5. 567
6. 1000

Build Your Understanding

Make a Math Creature

You Will Need
- felt pens
- coloured pencils
- white paper
- calculator

Work individually.

1. Choose a number.

2. Brainstorm many different math sentences for which your number is the answer. Use a calculator to check your math sentences.

3. Draw a creature that has many legs. On each leg of the creature, record one of the math sentences that you brainstormed.

4. Make a pattern by colouring in your creature.

5. Describe the pattern you created in your creature.

6. Think of a name for your creature.

Start-Up Math

What Did You Learn?

1. Share your math creature with a classmate.
2. Which operation did you find easiest to use when making questions for your creature? Why?
3. Which operation did you find most challenging to use when making questions for your creature? Why?
4. Is the name of your creature appropriate? Why or why not?
5. What could you change to improve your creature?
6. What lines of symmetry can you find in your creature? How do you know they are lines of symmetry?

Practice

1. Make up four questions for each answer below. One of the math sentences for each answer must use multiplication.

   a) 16   **b)** 21   **c)** 36   **d)** 56   **e)** 72

2. Answer the following:

   a) 16 + 23 = ■   **b)** 24 + 35 = ■

   c) 86 + 17 = ■   **d)** 69 + 79 = ■

   e) 146 + 257 = ■   **f)** 4563 + 3421 = ■

   g) 17 − 8 = ■   **h)** 36 − 15 = ■

   i) 75 − 26 = ■   **j)** 359 − 47 = ■

   k) 6675 − 301 = ■   **l)** 7504 − 2592 = ■

Extension

You Will Need
- base-ten blocks

3. Show each of the math sentences on your math creature using base-ten blocks. Have a partner check your work.

Lesson 3
Making Geometric Figures

Get Started

1. Draw a picture of each of these shapes:
   a) circle   b) triangle   c) rectangle   d) square
   e) parallelogram   f) rhombus   g) pentagon

2. With a small group, discuss what you know about each shape from question 1.

Build Your Understanding

Make Garden Figures

You Will Need
• pattern blocks

Work with a small group.

1. Choose a number between 10 and 100 to represent the value of each pattern-block shape. Keep a record of your decisions.

2. Each person in the group uses the pattern blocks to make a figure of something found in a garden.

3. Trace each figure into your notebook and then calculate its value.

4. Record your work by making a math sentence for each design.

5. Have a group member check your math sentence.

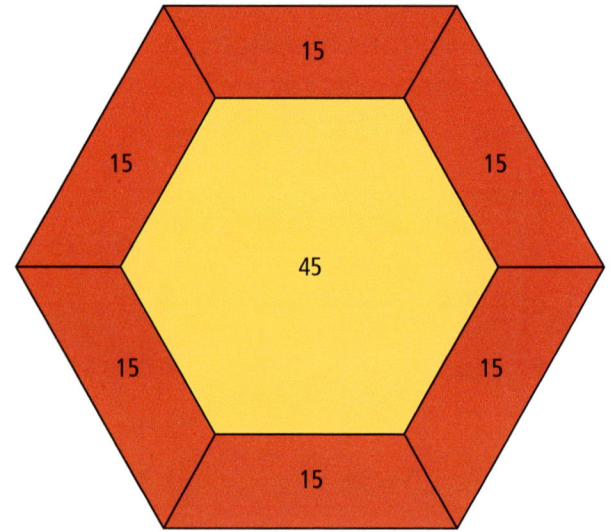

15 + 15 + 15 + 15 + 15 + 15 + 45 = 135

Start-Up Math

What Did You Learn?

1. Which garden figure in your group had the greatest value?
2. Which garden figure in your group had the least value?
3. Would a figure with six pattern-block pieces always have a greater value than a figure with four pieces? Explain.
4. Arrange the figures that your group created in order from least value to greatest value.
5. What is the total value of everything in your group's garden?

Practice

In your notebook, make a chart like this one. Check the boxes for which the statements are always true. Compare your answers with a classmate's.

Shape	Statement				
	Has Four Square Corners	Has Two Pairs of Parallel Sides	All Sides Are the Same Length	Has Four Sides	Opposite Sides Are the Same Length
square					
rectangle					
equilateral triangle					
rhombus					
trapezoid					
quadrilateral					
parallelogram					

Lesson 3: Making Geometric Figures

A Math Game to Play

Play the game Number Find with a partner.

How to Play

1. Take a turn by secretly choosing one of the numbers in this illustration.

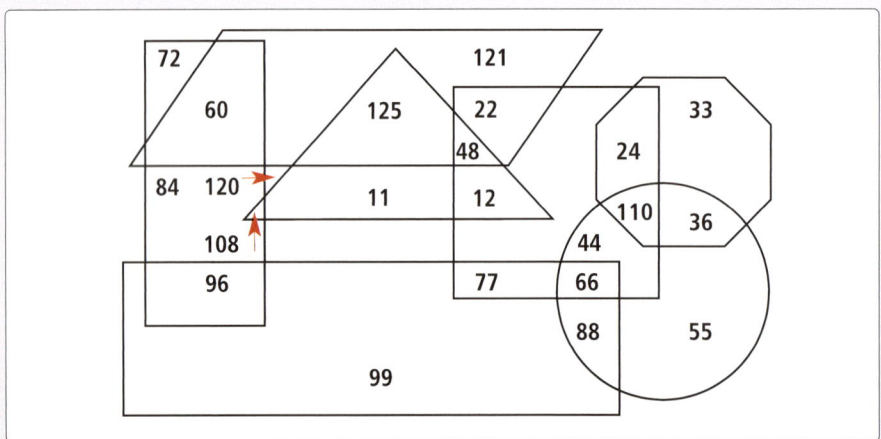

2. Your partner asks questions to guess your number. For example, "Is the number in the circle?"
3. You can only answer yes or no.
4. Your partner continues to ask questions until he or she finds your number.
5. Keep track of the number of questions asked.
6. Switch roles. This time your partner picks a secret number, and you ask the questions. Remember to keep track of the number of questions asked for each turn.
7. The person who asks the fewest questions in five turns is the winner.

MEASUREMENT

Lesson 4

Measuring With String and Cubes

Get Started

You Will Need
- 30 cm length of string
- cm ruler

1. Estimate then measure these lengths of string in centimetres.

   a)

   b)

   c)

   d)

   e)

2. How long is each length of string in millimetres?

Lesson 4: Measuring With String and Cubes

Build Your Understanding

Linking Cube Measures

You Will Need
- linking cubes (1 cm³ each)

Work individually.

1. Gather 50 linking cubes. Each linking cube is 1 cm³.
2. Use your linking cubes to measure 10 objects in your classroom.
3. Record your answers in a chart.
4. How many decimetres and millimetres would each object be?

What Did You Learn?

1. Share your work with a classmate.
2. How are centimetres and millimetres related?
3. How do you measure curved or round objects?
4. How could you use linking cubes to find the perimeter of a quadrilateral? Use pictures, numbers, and words to explain your answer.

Practice

1. Use 1-cm grid paper, a ruler, and string to draw the following:

    a) a pencil that is 17.5 cm long

    b) a key that is 61 mm long

    c) a ring that has a circumference of 4.1 cm

    d) a piece of chalk that is 7.8 cm long

    e) a triangle with a perimeter of 38 cm. Two sides are 12 cm each.

2. Use 1-cm grid paper to draw a square and find its perimeter. Explain how you got your answer.

GEOMETRY AND SPATIAL SENSE

Problems to Solve

Here are some fun problems for you to solve. For each problem, a helpful problem-solving strategy is included for you to use. Later in the year, once you have learned more strategies, you will have the chance to choose the strategies you want to use.

Problem 1

Math at the Movies

STRATEGY: ACT IT OUT

Playing a role or acting out a problem lets you see the problem more clearly and helps you figure out a solution.

OBJECTIVE:

Demonstrate an understanding of coordinate systems (rows and columns)

Problem-Solving Steps

There are four steps you can follow to help you solve a math problem. You will be reminded of these steps throughout the year:

1. **Understand the problem:** Rewrite the problem in your own words. If you can, draw a picture of the problem. List or highlight important numbers or words.
2. **Pick a strategy:** For example, "Act It Out," "Draw a Picture," "Use Objects," and "Guess and Check."
3. **Solve the problem:** Use a strategy to solve the problem. Describe all steps using math words and/or symbols. Try a different strategy if you need to. Organize the results using a diagram, model, chart, table, or graph.
4. **Share and reflect:** Did the strategy you picked work? Would a different strategy also work? Does your solution make sense? Could there be more than one answer to the problem? How did other people in your class solve the problem?

Problem

Imagine that a Grade 5 class is taking a trip to a movie theatre. The seats at the theatre are in rows and columns. When the students arrive at the theatre, Tomas sits two seats ahead of Kenny. Jamar is seated three seats to the right of Tomas. Jessica is sitting three seats behind Jamar. Shefali is seated two seats back and three to the left of Kenny. Where is Jessica seated in relation to Kenny? Work in a small group and use the "Act It Out" strategy to solve this problem.

Reflection

1. What information was given to help you solve the problem?
2. What did you need to figure out?
3. Do you think the "Act It Out" strategy was a good strategy to use? Why or why not?
4. What other strategies do you know that you could have used to solve this problem?

Extension

With your group, make up another problem that can be solved using the "Act It Out" strategy. Write down your problem, and give it to another group to solve.

Journal

Write what you know about the "Act It Out" strategy, or write about a time when you used the "Act It Out" strategy.

DATA MANAGEMENT AND PROBABILITY

Problem 2
Family Reunion Plan

STRATEGY: DRAW A PICTURE OR DIAGRAM

You can draw simple pictures or diagrams to help you solve a problem. Pictures or diagrams let you see how the parts of a problem work together.

OBJECTIVE:
Create a tree diagram

Problem

Every two years Grandma and Grandpa Fehr have a family reunion. They have 5 children. $\frac{2}{5}$ of their children are boys. Neither of their sons brings his wife, and neither has any children. 2 of Grandma and Grandpa Fehr's daughters bring their husbands.

1. If the daughters bring 6 of their children in total, how many people will be coming to the reunion altogether?

2. If the Fehrs use 3 tables for dinner, what type of seating arrangement could they use?

Use the "Draw a Picture or Diagram" strategy to solve this problem.

25 years ago

Problems to Solve

Reflection

1. What did you know about the families before you solved the problem?

2. What did you need to figure out?

3. Share your solution with another classmate. How are your solutions different, and how are they the same?

4. Do you think the "Draw a Picture or Diagram" strategy was a good strategy to use for this problem? Why or why not?

Extension

1. Compare the number of people at the Fehr family reunion to the number of people in the family picture shown on the previous page. Write three number sentences to explain the difference in family size.

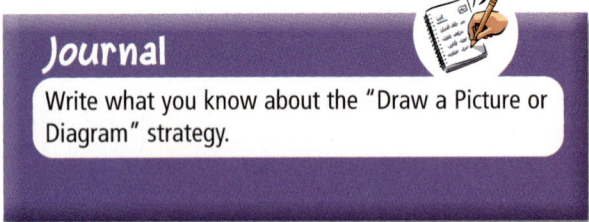

Journal

Write what you know about the "Draw a Picture or Diagram" strategy.

2. Write a similar math problem about your family. Ask a classmate to solve it.

Start-Up Math

NUMBER SENSE AND NUMERATION

Problem 3
Counting Animals

STRATEGY: USE OBJECTS
Using objects can help you organize information so you can see the solution. You can use simple objects, such as blocks or pieces of paper.

OBJECTIVE:
Solve problems involving whole numbers

Problem

You Will Need
- base-ten blocks

Willy the farmer had a total of 24 pigs and chickens on his farm. He wanted his granddaughter, Jane, to practise her counting, so Willy told her to count the legs of all the pigs and chickens. When she was done, she told him, "There are 72 legs."

1. How many pigs might the farmer have had?
2. How many chickens might the farmer have had?

Use the "Use Objects" strategy to solve this problem.

Reflection

1. What did you need to figure out to solve this problem?
2. Explain how you used objects to help you solve this problem.
3. Share your answers to the problem with another classmate. Are your answers different? Why?

Extension

Make this problem either easier or more challenging. Give your new problem to a classmate to solve.

Problems to Solve

PATTERNING AND ALGEBRA

Problem 4

Estimating and Checking on a Nature Walk

STRATEGY: GUESS AND CHECK

One way to solve a difficult problem is to make guesses and then check to see if your answers are correct. You can use this strategy when you are working with large numbers or if the problem involves many pieces of information. Sometimes what you find out as you use the "Guess and Check" strategy leads you to other strategies.

OBJECTIVE:

Estimate to find a pattern

Problem

For four days in a row, you go on a nature hike through a nearby forest. Each day, you pass the same spruce tree. You notice that cones have been dropping from it in the following pattern:

1 cone drops the first day.

2 cones drop the second day.

4 cones drop the third day.

8 cones drop the fourth day.

1. How many cones will have dropped on day 5? on day 10? on day 30?

2. Explain the pattern in which the cones fall. State the rule for the pattern.

Use the "Guess and Check" strategy to solve the problem and to find the pattern.

16 Start-Up Math

Reflection

1. Explain how you came up with the pattern.

2. How did the "Guess and Check" strategy help you find out how many cones dropped on day 5, day 10, and day 30?

3. Do you think this strategy was a good strategy to use? Why or why not?

Extension

Look at the leaves on a small bush, tree, or plant. Do you see any patterns?

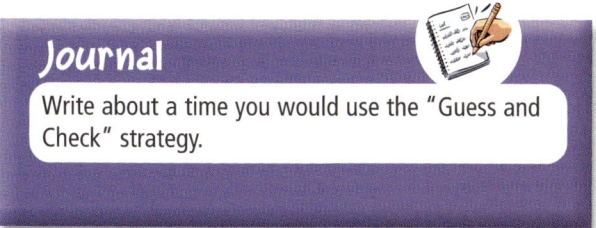

Journal

Write about a time you would use the "Guess and Check" strategy.

Problems to Solve

Problem 5
Backwards Bricklaying

STRATEGY: WORK BACKWARDS

Sometimes the best way to solve a problem is to begin with the answer or information at the end of the problem and work backwards toward the beginning.

OBJECTIVE:

Explain your thinking when solving a fraction problem

Problem

The 4 children are helping their parents by laying bricks to make a square driveway. The oldest child lays the most bricks, the twins each lay the same number of bricks, and the youngest child lays $\frac{1}{4}$ of the driveway, which consists of $205\frac{1}{2}$ bricks. Based on this information, figure out three different ways that the 4 children can share the work of laying the bricks. Use the "Work Backwards" strategy to solve this problem.

Tip
Start with the information that was presented last.

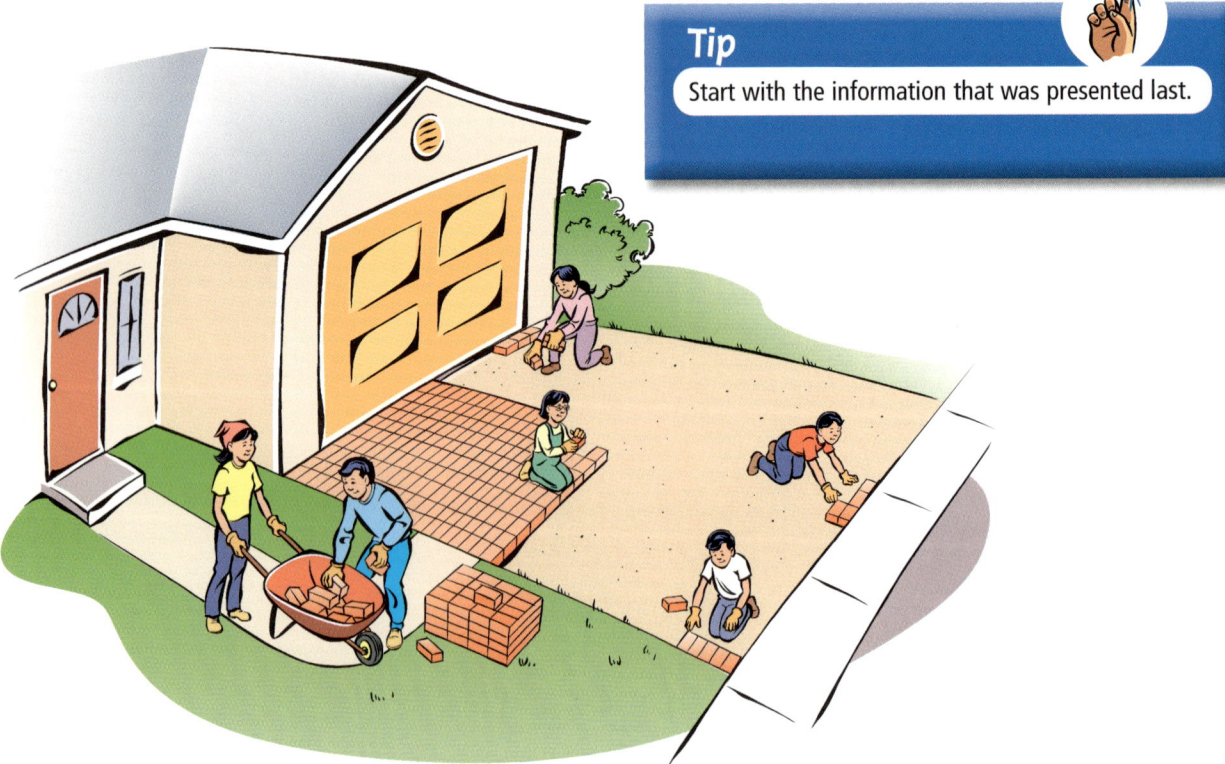

Reflection

1. What information did you know before you started to solve the problem?

2. What information did you need to figure out to solve the problem?

3. How did you figure out how many bricks there were in total?

4. Do you think the "Work Backwards" strategy was a good strategy to use? Why or why not?

5. Explain your solution to a classmate. Are your solutions the same or different? Explain why.

Extension

If the last child lays $278\frac{1}{4}$ bricks, which is still $\frac{1}{4}$ of all the bricks, how many bricks would be needed altogether?

Unit 1
The Mathematics of Weather

For hundreds of years, people have used math to explore weather. We use math to study and learn about rain, snow, wind, clouds, and temperature. We also use math to examine weather over periods of time to identify weather patterns.

In Chapter 1, you will develop your skills in operations with whole and decimal numbers so you can compare weather data. In the Chapter Wrap-Up, you will apply what you have learned to create a comparison chart or a poster of world, Canadian, and local weather records.

In Chapter 2, you will learn about the units used to measure weather. You will also practise collecting, organizing, presenting, and analyzing weather data. These important skills will prepare you for the Chapter Wrap-Up, where you will choose the best type of graph for presenting weather data.

In Chapter 3, you will identify and explain weather patterns. In the Chapter Wrap-Up, you will present a report on important weather patterns in your community.

Chapter 1

Weather Numbers

Mathematics allows us to measure and record weather events. This important information helps people better understand weather and long-term weather patterns, known as climate.

In this chapter, you will
- read and write whole numbers in standard, expanded, and written forms
- read, write, compare, and order decimal numbers
- compare and order whole numbers up to 100 000 and decimal numbers in the hundredths
- add, subtract, multiply, and divide whole numbers and decimal numbers
- count by elevens, then by twelves
- develop mental computation and problem-solving strategies and apply them to solve weather-related number problems

At the end of this chapter, you will work as part of a class weather-station team to complete a chart or a poster comparing weather information for the world, for Canada, and for your school community. You will then create and solve problems relating to your chart or poster.

Chapter 1: Numbers in Weather

Lesson 1

Exploring Place Value

WEATHER REPORT

PLAN:
You will develop your understanding of place value. You will then use this knowledge to compare and order numbers from whole numbers in tens of thousands to decimal numbers in the hundredths.

DESCRIPTION:
An anemometer is a weather instrument used to measure wind speed. Meteorologists measure wind speed in kilometres per hour (km/h).

An anemometer

Get Started

Place value tells you that each digit in a number has a different value, depending on its position. A place-value chart helps you read and understand numbers.

ten thousands	thousands	hundreds	tens	ones	tenths	hundredths
7	8	1	9	6 .	4	2

1. Work with a partner. Use the place-value chart above to read the number out loud. Take turns explaining what each digit in the number means.

2. How could you use your knowledge of place value to order the wind speeds (km/h) from greatest to least?

Highest Average Wind Speeds in Canadian Provinces		
Province	Wind Speed (km/h)	Place
Manitoba	22.7	Churchill
Saskatchewan	22.9	Swift Current

Chapter 1: Weather Numbers

3. What is the place value of each underlined digit?
   a) <u>2</u>2.9   b) 2<u>2</u>.9   c) 22.<u>9</u>
4. Which number is greater, 22.7 or 22.9? How do you know?
5. Explain what strategies you use when deciding which number is greater.

Vocabulary

digit: A symbol used to record a number. 0, 1, 2, 3, 4, 5, 6, 7, 8, and 9 are digits.

place value: The value given to the place in which a digit appears in a number. For example, in 12 683, 1 is in the ten-thousands place, 2 is in the thousands place, 6 is in the hundreds place, 8 is in the tens place, and 3 is in the ones place.

Build Your Understanding

Use Place Value to Compare and Order Numbers

1. This chart shows the highest average wind speed in Canadian provinces and territories. Arrange the average wind speeds from greatest to least.

Highest Average Wind Speed		
Province or Territory	Wind Speed (km/h)	Place
Newfoundland and Labrador	28.0	Bonavista
Prince Edward Island	22.4	Summerside
New Brunswick	22.4	Miscou Island
Nova Scotia	25.7	Sable Island
Québec	32.0	Grindstone Island
Ontario	21.0	Bruce Power Plant
Manitoba	22.7	Churchill
Saskatchewan	22.9	Swift Current
Alberta	21.5	Pincher Creek
British Columbia	33.7	Cape St. James
Yukon	14.1	Whitehorse
Northwest Territories	19.9	Nicholson Peninsula
Nunavut	35.3	Resolution Island

Source: Environment Canada

2. Meet with a classmate to share, compare, and discuss your results for question 1. Explain how knowing about place value helped you arrange the numbers correctly.

Lesson 1: Exploring Place Value

What Did You Learn?

1. Write a definition of "place value" in your own words. Use examples to support your definition.

2. Explain to a classmate the strategy you will use to decide which of the following numbers is greater: 13 456.71 or 13 456.73.

3. What patterns did you notice when you organized average wind speeds from the greatest to least?

Practice

Vocabulary

\>: A symbol meaning "is greater than," as in 9.6 > 9.4
<: A symbol meaning "is less than," as in 8.48 < 8.49

Make these statements true. Use < or >.

1. 541 ■ 642
2. 76 ■ 78
3. 9322 ■ 9232
4. 87 311 ■ 78 131
5. 35.3 ■ 33.5
6. 19.95 ■ 20.01
7. 15.31 ■ 15.32

8. Arrange these numbers in order from greatest to least.

   578   785   758   857   875   587

A Math Problem to Solve

Scientists developed a way of classifying hurricanes that gives people a good idea of what to expect from a storm. This classification system is called the Saffir-Simpson scale. Hurricanes are ranked from 1 (least dangerous) to 5 (extremely dangerous).

Category	Wind Speed (km/h)	Storm Surge* (metres)	Description
1	118–153	1.2–1.7	weak
2	154–177	1.8–2.7	moderate
3	178–210	2.8–3.9	strong
4	211–249	4.0–5.5	very strong
5	> 250	> 5.5	devastating

*Storm surge is the height that ocean water reaches above the normal tide.

9. A hurricane has a wind speed of 243 km/h. What category is it?

10. The wind speed of the hurricane drops 50 km/h. Will this drop change the category of the storm? If so, what will the new storm category be?

11. Estimate the average storm surge of the weakened hurricane, and explain how you made your estimate.

12. How did an understanding of place value help you read, understand, and compare weather numbers?

Technology

Conduct an Internet search on the windiest places in the world. Record your results on a spreadsheet, and display the results in a bar graph. Use the text tool to make some statements about your data.

Lesson 2

Exploring Whole Numbers

WEATHER REPORT

PLAN:
You will learn about whole numbers and how to read and write whole numbers in standard, expanded, and written forms.

DESCRIPTION:
Numbers are important to help us measure, record, describe, and understand weather.

The rainstorm that caused the 1996 Saguenay flood dropped an average of nearly 126 mm of rain over a 100 000 km² area in 48 h.

Get Started

Whole numbers are the set of numbers 0, 1, 2, 3, 4, 5, 6, and so on. These numbers do not contain any fractions. The set continues indefinitely.

When and how are whole numbers used to talk about the weather? Share your answers. Compile a class list of ideas that you can add to throughout the unit.

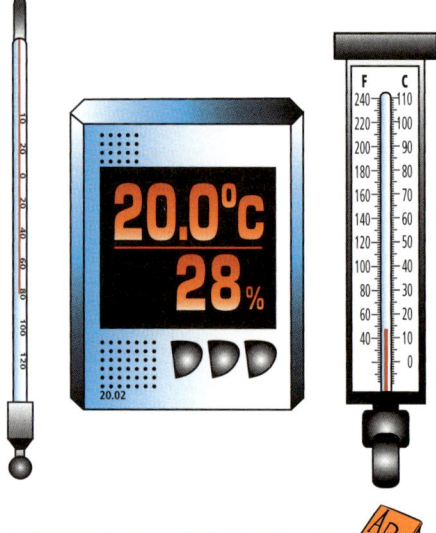

Whole numbers can be written in different ways. Let's look closely at the whole number 3789.

- In standard form, the number is 3789.
- In expanded form, 3789 is 3000 + 700 + 80 + 9.
- In written form, 3789 is three thousand seven hundred eighty-nine.

Vocabulary

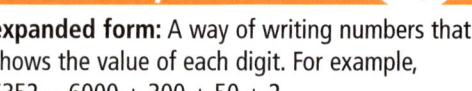

expanded form: A way of writing numbers that shows the value of each digit. For example, 6352 = 6000 + 300 + 50 + 2.

standard form: A way of writing numbers in which each digit has a place value according to its position in relation to other digits; for example, 6352

written form: A way of writing numbers in words. For example, 6352 in written form is "six thousand three hundred fifty-two."

Build Your Understanding

Read and Write Whole Numbers in Standard, Expanded, and Written Forms

1. In your notebook, create and complete a chart like this one.

	Whole Numbers		
Drawing	Standard Form	Expanded Form	Written Form
(drawing of sticks)			
			ninety-nine thousand four hundred seventy-one
	80 156		
		70 000 + 7000 + 500 + 30 + 9	
			one hundred thousand

2. Work with a classmate. Take turns reading out loud all the standard and written forms of the whole numbers in each row of the chart. Listen carefully as your partner reads; then help him or her correct any mistakes.

3. Use a separate piece of paper. Create a chart like the one above, with different whole-number examples.

4. Exchange charts with a classmate and challenge him or her to complete yours. Check your partner's answers to make sure they are correct.

Lesson 2: Exploring Whole Numbers

What Did You Learn?

1. In your own words, define "whole number." Provide an example of a six-digit number to go with your definition. Make a drawing of your example that clearly shows the value of each digit.

2. Write 30 000 + 7000 + 800 + 20 + 3 in standard form.

3. Write 45 670 in expanded form. How can knowing the expanded form of a number help you read and write it?

4. Use written form to write 90 643.

Journal

Set up a two-column chart like the one below. It will be used to compare the different types of numbers you will be learning about in this chapter. Title your chart "Types of Numbers." In your own words, fill in a definition for whole number. You will complete the other chart heading and column later on in the chapter.

Whole Number	

Practice

Write the following whole numbers in standard form.

1. 40 000 + 6000 + 900 + 70 + 9
2. 15 000 + 300 + 40 + 7
3. eighty-two thousand four hundred nineteen

Write the following whole numbers in expanded form.

4. 7353
5. fifty-five thousand fifty-one
6. 89 776
7.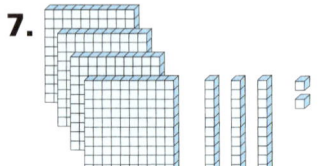

Journal

Make a note about any type of whole number you have difficulty reading or writing. Jot down a strategy or strategies that might help you read or write those problem numbers.

Write the following whole numbers in written form.

8. 11 101
9. 40 000 + 700 + 90 + 6
10. 66 282
11. 99 999
12. Work with a partner to take turns reading out loud all the whole numbers in questions 1 through 11.

NUMBER SENSE AND NUMERATION

Lesson 3
Exploring Decimal Numbers

WEATHER REPORT

PLAN:
You will learn to read, write, compare, and order decimal numbers.

DESCRIPTION:
Precipitation can be rain, snow, sleet, or hail. Meteorologists, people who study weather, describe "light rain" as less than 0.5 mm per hour and "heavy rain" as more than 4 mm per hour. Meteorologists use gauges to measure amounts of precipitation.

Get Started

Amounts of precipitation can be written using decimal numbers. A fraction that has a denominator of 10, 100, 1000, and so on, can be written as a decimal number.

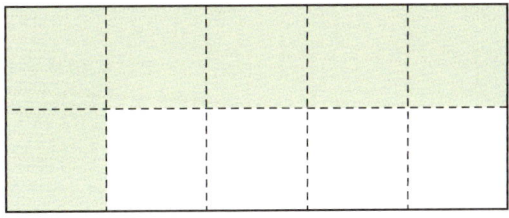

$\frac{6}{10}$ or 0.6

Lesson 3: Exploring Decimal Numbers 29

The chart below will help you understand and read decimal numbers.

Decimal Numbers		
Fraction	Decimal	How to Read the Decimal Number
$\frac{1}{10}$	0.1	one tenth
$\frac{3}{10}$	0.3	three tenths
$\frac{1}{100}$	0.01	one hundredth
$\frac{3}{100}$	0.03	three hundredths

1. Work with a partner. Take turns reading the decimal numbers out loud.

2. Use the pattern in the chart to help you do the following:
   a) Write $\frac{7}{10}$ as a decimal number, then read the decimal number out loud to your partner.
   b) Write $\frac{7}{100}$ as a decimal number, then read the decimal number out loud to your partner.

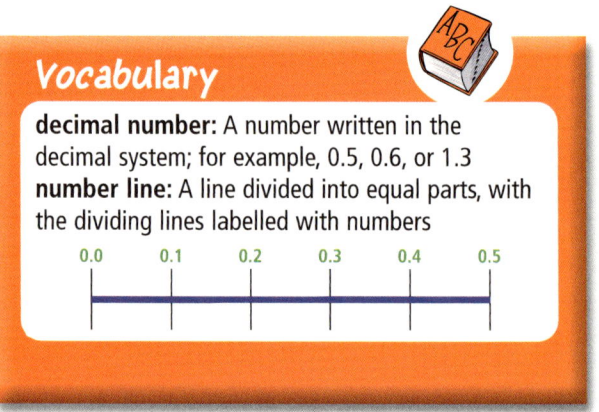

Vocabulary

decimal number: A number written in the decimal system; for example, 0.5, 0.6, or 1.3
number line: A line divided into equal parts, with the dividing lines labelled with numbers

3. How many hundredths are there in $\frac{3}{10}$? Explain your reasoning.

4. Read the number 83 492.73 using the place-value chart below to help you.

hundred thousands	ten thousands	thousands	hundreds	tens	ones	.	tenths	hundredths

5. Make a list of fractions that have a denominator of 10 or 100. Have a classmate write each of the fractions on your list as a decimal number; then read the decimal number out loud. Make sure you agree on the answers.

Technology

Explore decimals using a calculator. Begin with a single digit number (such as 5), and note where the decimal place is. Divide this number by 10, 100, and 1000. Discuss what happens. Predict answers for another single-digit number. Check your predictions. Then try using a two-digit number to begin. Do your results change? Why or why not? Share and compare your findings with a partner.

Chapter 1: Weather Numbers

Build Your Understanding

Read, Write, and Order Decimal Numbers

1. Kenny started recording the amount of precipitation in his community over five different days. Copy and complete Kenny's chart in your notebook.

Amount of Precipitation (mm)		
Fraction	Decimal	How to Read the Decimal Number
$\frac{5}{10}$		
	0.02	
		eighteen hundredths
		forty hundredths
$\frac{9}{10}$		

2. For each decimal number in the chart, draw a picture to show its value. For example, in Get Started, a part of a rectangle strip was shaded to show that it represents the decimal 0.6. Be creative.

3. Make a chart like the one above. Fill in one cell in each row of the chart. (Fill in a different cell in each row to make your chart more challenging.) Exchange unfinished charts with a classmate, and challenge him or her to complete yours. Check each other's answers.

What Did You Learn?

1. Which of the following are decimal numbers? Explain why.
   a) $\frac{3}{8}$ b) 7   c) $1\frac{3}{4}$ d) 0.45   e) 1.17

2. Can a whole number also be a decimal number? Explain.

3. In your own words, write a definition of a "decimal number." Give three examples of decimal numbers to support your definition.

4. Arrange your examples from question 3 in order from least to greatest. Explain to a classmate how you decided which decimal number was greatest.

Lesson 3: Exploring Decimal Numbers

Practice

1. Arrange the following numbers in order from least to greatest:
   a) 0.5
   b) seventy-five hundredths
   c) $\frac{4}{10}$
   d) 0.04

2. Explain how you decided on your order in question 1.

Complete the following sentences in your notebook:

3. 4 tenths is the same as ■ hundredths.

4. 50 hundredths is the same as ■ tenths.

5. 10 tenths is the same as ■ hundredths, or ■.

6. 1 is the same as ■ tenths and ■ hundredths.

7. Explain why 7 tenths and 70 hundredths are equivalent. Use pictures, numbers, and words.

Extension

8. Make a rain gauge and record rainfall.

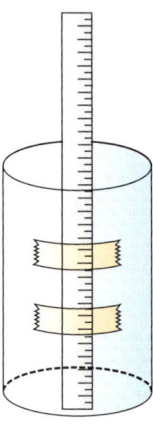

You Will Need
- straight-sided glass container
- cm ruler
- clear sticky tape

   a) Work with a partner to make a rain gauge like the one shown in the picture.

   b) Starting on a Monday, record the precipitation each day for five days. Record measurements in millimetres. If possible, use decimals in your readings.

   c) Each day, compare your readings with precipitation reported in local newspapers or on radio, television, or reliable Internet sites.

   d) If there is no precipitation one day, continue to record the precipitation until you have five days of data.

   e) Use the data you collected over the five days to create a bar graph of your findings. Remember to label your graph clearly.

f) Meet with another pair of students to share, compare, and discuss your data and graphs. How did you use decimal numbers in the activity? What patterns did you find in your precipitation data? What conclusions can you draw from these patterns?

g) If you were to do this activity again, how might you do it differently? Explain.

Journal

In Lesson 2, you started a chart titled "Types of Numbers" and wrote a definition of a whole number. Add the heading "Decimal Number" to your chart and enter a definition for it.

Whole Number	Decimal Number

Show What You Know

Review: Lessons 1 to 3, Whole Numbers and Decimal Numbers

1. Write the following numbers in expanded form:

a) 61 080   b) 9.85
c) 30.94   d) 2804
e) 71.86   f) 48 810

2. Write the following numbers in written form:

a) 90 542   b) 29.80
c) 4.07   d) 11.18
e) 7891   f) 2980

3. Change the following fractions to decimals:

a) $\frac{8}{100}$ b) $\frac{5}{10}$

c) $\frac{49}{100}$ d) $\frac{75}{100}$

e) $\frac{11}{100}$ f) $\frac{2}{100}$

4. Order the following numbers from greatest to least:

23   22.25   23.10   22.09   22.24

NUMBER SENSE AND NUMERATION

Lesson 4

Exploring Number Patterns

WEATHER REPORT

PLAN:
You will count by elevens and then by twelves to develop your counting skills.

DESCRIPTION:
Lightning strikes the earth 50 to 100 times every second. Toronto's CN Tower is struck by lightning about 30 times each year during a lightning season that can run from April to October.

Get Started

In the above description, you could say that lightning strikes the CN Tower "tens of times" during a season. The expression "tens of times" tells you to group or count by tens.

1. With a partner, take turns counting by tens to 100. How many "tens of times" are there in 50? 60? 70? 100?

2. Draw a number line that will help you count by nines to 108. With a partner, use your number line to count to 108 out loud.

> **Journal**
> Make notes on any useful counting strategies and aids you know. For example, you might find a number line helpful when counting by elevens and twelves. What other counting aids could you use?

Chapter 1: Weather Numbers

Build Your Understanding

Count by Larger Numbers

You Will Need
- blank times-table chart

1. Complete the times-table chart.

2. Use the times-table chart to complete these patterns.
   a) 4, 8, 12, 16, … 48
   b) 5, 10, 15, 20, … 60
   c) 6, 12, … 72
   d) 7, 14, … 84
   e) 8, 16, … 96
   f) 11, 22, 33, …132
   g) 12, 24, 36, … 144

> **Tip**
> When you compare numbers, work from left to right—look at thousands first, then hundreds, then tens, and finally ones.

3. Work with a partner to take turns counting the following out loud:
   a) by elevens to 132   b) by twelves to 144

4. Counting is very important when estimating the location of lightning in relation to a person. Thunder is the sound of lightning moving rapidly through the atmosphere. To find out roughly how close lightning is to where you are, count the seconds between the lightning flash and the sound of thunder. Each second represents 300 m.

   a) Jamar counts 30 s between a lightning flash and thunder. About how far away is the lightning storm? Express your answer in metres and then in kilometres.

   b) Jessica is travelling in another part of the province and counts 4 s between the lighting flash and thunder. How far away from the lightning is she?

> **Journal**
> Explain the strategy you used to solve these problems. Which number operations did you use? Why? Did counting strategies help you? If so, how?

Lesson 4: Exploring Number Patterns

What Did You Learn?

1. How high can you mentally count by elevens?
2. How high by twelves?
3. How might you improve your counting skills?
4. Where and how do you apply counting skills in everyday life?
5. How is counting by elevens and twelves similar to multiplying by these numbers?
6. What strategy or strategies can you use to help you count by elevens and twelves?

Practice

Mentally multiply these numbers. Record the completed equations.

1. $11 \times 6 =$ ■
2. $12 \times 4 =$ ■
3. $6 \times 12 =$ ■
4. $8 \times 11 =$ ■
5. $9 \times 11 =$ ■
6. $12 \times 9 =$ ■
7. $3 \times 12 =$ ■
8. $11 \times 12 =$ ■

Technology

Research what to do if you are caught in a thunderstorm, using on-line or CD-ROM resources. Create a poster about lightning safety, using the information you gathered.

Find the missing numbers. Record the completed equations.

9. ■ $\times 12 = 144$
10. ■ $\times 11 = 22$
11. $12 \times$ ■ $= 60$
12. $11 \times$ ■ $= 33$
13. ■ $\times 11 = 11$
14. $12 \times$ ■ $= 120$

15. Record multiples of 11 until you get to 132. What patterns do you see?
16. How can you use the 10 times table to help you multiply by 11? by 12?
17. Record multiples of 12 until you get to 144. Explain how you could use the 6 times table to help you multiply by 12.
18. Create flash cards that will help you practise your 11 and 12 times tables.

Lesson 5
Adding and Subtracting Whole Numbers

WEATHER REPORT

PLAN:
You will practise adding 3 four-digit numbers and subtracting 2 four-digit numbers. As well, you will solve weather-related problems requiring addition and subtraction of whole numbers.

DESCRIPTION:
People who study climate and weather in Canada keep records of the hours of sunlight and the amount of precipitation in communities each year.

Estevan, Saskatchewan

Prince Rupert, British Columbia

Get Started

For each addition and subtraction question,
- estimate the sum or difference
- calculate the answer
- check the reasonableness of your calculation by comparing the estimate with the answer
- double-check your calculations with a calculator

Look at the addition and subtraction examples on the next page to help you.

Addition

Addition Example Question

Add the ones	Add the tens	Add the hundreds	Add the thousands
2 trade 20	2 2 trade 200	2 2 2 trade 2000	2 2 2
1885	1885	1885	1885
1497	1497	1497	1497
+ 1818	+ 1818	+ 1818	+ 1818
0	00	200	5200

1. a) 2763
   1499
   + 871

   b) 4566
   3686
   + 523

   c) 9484
   7263
   + 1438

   d) 5883
   5822
   + 5656

2. Estevan, Saskatchewan, receives the most hours of sunshine in Canada each year, at about 2500 h. Regina, Saskatchewan, receives about 2333 h of sunshine, and Churchill, Manitoba, receives 1828 h. Altogether, how many hours of sunshine do these three communities receive in an average year? Show your work.

Subtraction

Subtraction Example Question

Subtract the ones	Subtract the tens	Subtract the hundreds	Subtract the thousands
0 13 trade 10	4 10 13 trade 100	4 10 13 no trading needed	4 10 13
1513	1513	1513	1513
− 1169	− 1169	− 1169	− 1169
4	44	344	344

3. a) 3363
   − 1251

   b) 5492
   − 2703

   c) 8532
   − 4677

   d) 9111
   − 1999

4. Prince Rupert, British Columbia, is the wettest city in Canada, with a total annual precipitation of about 2552 mm. The annual precipitation in Vancouver is about 1113 mm. What is the difference between the annual precipitation in Prince Rupert and that in Vancouver? Show your work.

"Signal" Words Telling Which Operation to Use in a Problem	
Addition	Subtraction
sum	difference
total	how much more
both	how much less
altogether	decreased by

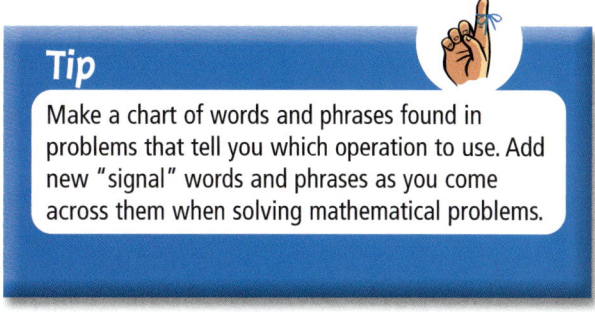

Tip

Make a chart of words and phrases found in problems that tell you which operation to use. Add new "signal" words and phrases as you come across them when solving mathematical problems.

5. For problems 2 and 4, write the words or phrases that told you which operation (addition or subtraction) to use.

Build Your Understanding

Solve Problems With Addition and Subtraction

Addition

1. Halifax, Nova Scotia, has an average of 1885 annual hours of bright sunshine. In St. John's, Newfoundland and Labrador, the average is about 1497 h, and in Charlottetown, Prince Edward Island, it is about 1818 h. What is the total annual number of hours of bright sunshine in these three communities?

2. Round each of the three numbers you added to the nearest thousand. Add the rounded numbers to check whether the sum is reasonable.

3. Check your calculations with a calculator.

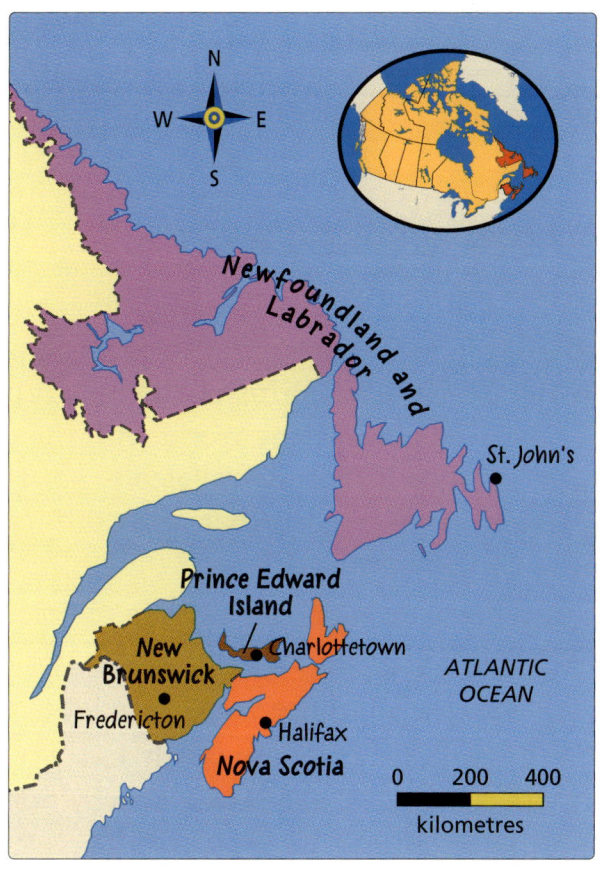

Lesson 5: Adding and Subtracting Whole Numbers

Subtraction

4. St. John's, Newfoundland and Labrador, receives about 1513 mm of precipitation each year. Charlottetown, Prince Edward Island, receives about 1169 mm. How much more precipitation does St. John's receive on average each year than Charlottetown?

5. Round each of the two numbers you subtracted to the nearest hundred. Subtract the smaller rounded number from the larger one to see if the difference is reasonable. Was the difference reasonable? Explain why or why not.

6. Check your calculations with a calculator.

7. Meet with a classmate or classmates and discuss why understanding place value is important when adding and subtracting whole numbers. Share your answers with the class.

Journal

Write in your own words the rules for finding the sum in an addition question, then for finding the difference in a subtraction question.

Vocabulary

rounding: Giving an approximate number for an exact number to a nearest place-value position. For example, 5769 to the nearest hundred is 5800.

What Did You Learn?

The place in Canada with the greatest average annual precipitation is Henderson Lake, British Columbia. It gets an average of 6655 mm a year. The place in the world with the greatest average annual precipitation is Mount Wailaleale, Kauai, Hawaii. It gets an average of 11 684 mm a year. What is the difference between the world and Canadian records?

1. What operation will you use to solve the problem? Why? What word or phrase tells you to use this operation?

2. Estimate what the solution will be; then calculate the answer.

3. Compare your answer with your estimate. Does it seem reasonable? Explain why or why not.

Practice

Write these whole numbers in standard, expanded, and written forms:

1. 25 000
2. 55 200
3. 6505
4. 11 684

> **Technology**
>
> Use on-line statistical information Web sites (such as www.statcan.ca *or* www.guinnessworldrecords.com) to gather data about weather. Use a word processing program to create some addition and subtraction problems from the data you collected. Challenge classmates to solve your problems.

Answer these addition or subtraction questions:

5.   2384
       1765
     + 417
	6. 7563
	       6812
	    + 4556
		7. 6783
		       4213
		      + 14
			8. 1637
			        208
			    + 1412

9.   7314
       6126
      + 10
	10. 1013
	        2987
	     + 1412
		11. 6018
		        5123
		      + 161
			12. 3612
			         307
			     + 1213

13.   6215
        316
     + 412
	14. 1081
	        2374
	     + 1016
		15. 6333
		        1423
		     + 2734
			16. 6412
			        512
			     + 2501

17.   7891
     − 997
	18. 6034
	    − 3149
		19. 6314
		     − 731
			20. 5621
			    − 1347

21.   9073
    − 4162
	22. 7315
	     − 402
		23. 2106
		    − 1203
			24. 6509
			     − 610

25.   5612
     − 428
	26. 8074
	     − 472
		27. 6372
		     − 419
			28. 6801
			    − 4230

Lesson 5: Adding and Subtracting Whole Numbers

NUMBER SENSE AND NUMERATION

Lesson 6

Multiplying and Dividing Whole Numbers

WEATHER REPORT

PLAN:
You will learn how to multiply two-digit numbers by two-digit numbers and divide four-digit numbers by one-digit numbers. Also, you will solve weather-related problems that require multiplication and division of whole numbers.

DESCRIPTION:
People who study weather keep records over long periods of time. They record such information as the number of tornadoes an area has or the amount of rain or snow it receives. Long-term weather data (information) helps us understand weather patterns and climate.

Kevin Shaw amidst the ruins of his home in Barrie, ON, on June 2, 1985. It was levelled by a tornado.

A street in Montréal under a heavy blanket of snow

Get Started

Knowing multiplication and division operations can help you understand weather events and patterns.

Multiplication

Giselle learned that Ontario has, on average, about 25 tornadoes a year. She wondered how many tornadoes this would be over 23 years.

To find the product, write the larger number above the smaller one.

The product is 575. Based on the average of 25 tornadoes a year over 23 years, Ontario would have 575 tornadoes in 23 years.

25 x 2**3** ――― 75	Multiply 25 x 3.
25 x **2**3 ――― 75 + 500 ――― 575	Multiply 25 x 20. Don't forget to move one space to the left since 2 is in the tens place. Add the partial products.

42   Chapter 1: Weather Numbers

Division

Tom looked at a chart of data on long-term weather averages. He found that the average annual precipitation in Sept-Îles, Québec, was 1125 mm. In Montréal, it was 946 mm, in Toronto 761 mm, and in Winnipeg it was 526 mm. The total mean annual precipitation in the four communities is 3358 mm. To find the mean yearly precipitation for each community, Tom must divide by 4. Why?

The quotient is 839 with a remainder of 2.

Use your knowledge of multiplication and division facts to decide if your answer is reasonable.

1. Round 3358 to the nearest hundred that 4 will divide into evenly.
2. How many times does 4 divide into this rounded number?
3. Is your answer in the original division problem reasonable? Explain why or why not.
4. Check your answer by multiplying the quotient by the divisor and adding the remainder.
5. Double-check your answer with a calculator.

```
     8
4 )3358
    32
```
Will 4 go into 33?
Yes, 8 x 4 = 32.

```
     8
4 )3358
  - 32
    15
```
Subtract and bring down the next number.

```
     83
4 )3358
  - 32
    15
  - 12
```
Will 4 go into 15?
Yes, 3 x 4 = 12.

```
     839
4 )3358
  - 32
    15
  - 12
    38
  - 36
     2
```
Subtract and bring down the next number.
Will 4 go into 38?
Yes, 9 x 4 = 36.
Subtract.
Since there is no number to bring down, and 4 will not go into 2, the remainder is 2.

Vocabulary

composite number: A number that has factors besides 1 and itself. For example, the number 6 has four factors: 1, 2, 3, and 6.
dividend: The number being divided in a division question; for example, in 36 ÷ 6, the dividend is 36
divisor: The number you are dividing the dividend by; for example, in 15 ÷ 3, the divisor is 3

factor: A number that divides evenly into another number. For example, 3 is a factor of 6.
prime number: Any number other than 1 whose only factors are 1 and itself. For example, 7 is a prime number.
product: The answer in a multiplication calculation
quotient: The answer in a division question
72 ÷ 9 = 8 ← quotient
   ↑   ↖ divisor
dividend

Lesson 6: Multiplying and Dividing Whole Numbers   43

Build Your Understanding

Multiply and Divide

1. Calculate the product for each of the following:

   a) 79   b) 36   c) 98   d) 57
      x 56   x 67   x 78   x 29

   e) 48   f) 78   g) 96   h) 89
      x 35   x 26   x 51   x 19

   i) 57   j) 37   k) 42   l) 49
      x 19   x 23   x 32   x 36

Tip
Estimate your product for each question. Use your estimate to decide whether or not your product is reasonable. Then, use a calculator to check the accuracy of your calculations.

2. Calculate the quotient and remainder for each of the following:

   a) 5897 ÷ 4   b) 3261 ÷ 5   c) 7890 ÷ 6
   d) 9344 ÷ 7   e) 7777 ÷ 8   f) 9874 ÷ 4

3. Round the dividend (the number being divided by) to the nearest number the divisor will divide into evenly. Use this estimate to decide whether or not your answer is reasonable.

4. Check your answers in each division question by using multiplication and, if there is a remainder, addition.

Journal
Write in your own words the algorithm, or the set of rules, for finding the product in a two-digit by two-digit multiplication question. Then write the algorithm for finding the quotient in a four-digit by one-digit division question.

Chapter 1: Weather Numbers

What Did You Learn?

1. Use one of the questions you completed in this lesson to explain to a classmate how multiplication and addition are related.

2. Use one of the questions you completed in this lesson to explain to a classmate how division is related to multiplication, subtraction, and addition.

3. Explain how you used estimation to check whether your answers were reasonable when you multiplied and divided.

Practice

Multiply the following:

| 1. 70 × 29 | 2. 45 × 41 | 3. 39 × 18 | 4. 47 × 10 | 5. 27 × 22 | 6. 54 × 30 |

Divide the following:

7. 42 ÷ 21   8. 68 ÷ 24   9. 95 ÷ 32   10. 86 ÷ 17   11. 75 ÷ 15
12. 6580 ÷ 3   13. 9435 ÷ 7   14. 5555 ÷ 5   15. 2070 ÷ 9

Math Problems to Solve

16. In Ontario, there is an average of 25 tornadoes a year. In Texas, there is an average of 125 yearly. How many times more tornadoes are there in Texas than in Ontario? Explain how you got your answer. Use pictures, numbers, and words.

17. In Penticton, British Columbia, the average annual precipitation is 283 mm a year. About how much is this per month? Show your work.

Show What You Know

Review: Lessons 4 to 6, Operations With Whole Numbers

1. With a partner, create a game that will help you practise your 11 and 12 times tables. Play your game.
2. Add or subtract: a) 5002 − 886   b) 4924 + 925 + 7461
3. Multiply or divide: a) 38 × 27   b) 8407 ÷ 9

NUMBER SENSE AND NUMERATION

Lesson 7
Adding and Subtracting Decimal Numbers

WEATHER REPORT

PLAN:
You will learn how to add and subtract decimal numbers to the hundredths place.

DESCRIPTION:
Snow is another type of precipitation. People who study weather and climate record snowfall amounts in decimal numbers. On television or in newspapers, you often see snowfall amounts expressed in centimetres.

Get Started

Add Decimal Numbers

Mr. Vavougis is concerned about the snow piling up on his roof. On the first day of the snowstorm, his town had 31.56 cm of snow. On the second day, 27.98 cm of snow fell. How much snow fell during the two days?

| hundred thousands | ten thousands | thousands | hundreds | tens | ones . | tenths | hundredths |

When adding decimal numbers, first arrange the numbers in a column, with the digits lined up according to their place values.

```
  31.56
+ 27.98
```

Add the hundredths
```
   1
  31.56
+ 27.98
  ─────
      4
```

Add the tenths
```
  1 1
  31.56
+ 27.98
  ─────
    .54
```

Add the ones
```
   1
  31.56
+ 27.98
  ─────
   9.54
```

Add the tens
```
  31.56
+ 27.98
  ─────
  59.54
```

59.54 cm of snow had fallen during the two days.

1. Round the numbers you added to the nearest ones place. Then, add mentally to estimate the approximate sum. Is the answer 59.54 reasonable? Explain why or why not.

2. Check your calculation with a calculator.

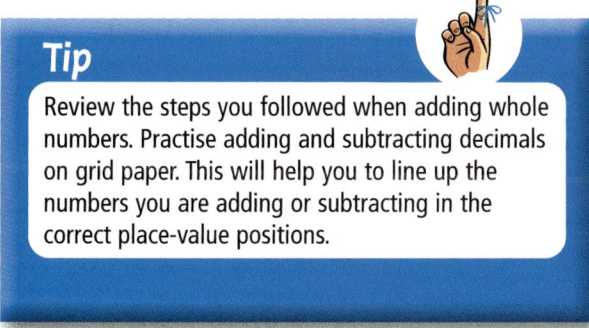

Tip
Review the steps you followed when adding whole numbers. Practise adding and subtracting decimals on grid paper. This will help you to line up the numbers you are adding or subtracting in the correct place-value positions.

Lesson 7: Adding and Subtracting Decimal Numbers

Subtract Decimal Numbers

At Mount Fidelity in British Columbia, 1.43 m of snow fell in one year. The same year, 0.69 m of snow fell at Mount Logan in Québec. What is the difference between the amount of snow that fell at Mount Fidelity and the amount that fell at Mount Logan?

When subtracting decimal numbers, place the larger number on top and arrange the digits according to their place value.

    1.43
− 0.69

Subtract the hundredths

   3 13
  1.4̸3   trade
− 0.69 hundredths
      4

Subtract the tenths

 0 13 13
 1̸.4̸3   trade
− 0.69 tenths
  .74

Subtract the ones

 0 13 13
 1̸.4̸3
− 0.69
 0.74

The difference between these annual average snowfalls is 0.74 m.

3. Round each of the numbers you subtracted to the nearest tenth. Then, subtract mentally to estimate the approximate difference. Is the answer 0.74 reasonable? Explain why or why not.

4. Check your calculation with a calculator.

Journal
When planning films, filmmakers draw storyboards to show each important step in a story. For example, in a how-to film about adding decimals, the first storyboard frame might show a student arranging numbers in the correct place-value positions. Draw storyboards for how-to videos on adding and subtracting decimals.

Technology
Using a spreadsheet with 1-cm square cells, input the numbers from the addition and subtraction questions in Get Started and Build Your Understanding. Remember to line up the decimal points properly.

Build Your Understanding

Add and Subtract Decimal Numbers

1. Add the decimal numbers.

   a) 454.08   b) 1001.20   c)   5.75   d)   5.78
         5.12   1.90   15.06   10.08
       + 0.63   + 25.05   + 0.35   + 116.62
       ——————   ——————   ——————   ——————

2. Arrange these numbers in columns in the correct place-value positions, and then add.

   a) 5.42 + 99 483.71 + 1.09   b) 711.02 + 715.32 + 0.11
   c) 11 872.5 + 9.43 + 7398.01 d) 42 + 796.30 + 444.71

3. Subtract the decimal numbers.

   a) 12.00   b) 107.49   c) 39.00   d) 19.45
      − 3.26   − 88.52   − 0.01   − 11.67
      ——————   ——————   ——————   ——————

4. Arrange these numbers in columns in the correct place-value positions, and then subtract.

   a) 10 − 2.37   b) 2.44 − 0.74   c) 11.29 − 0.42   d) 0.88 − 0.8

What Did You Learn?

1. Compare adding whole numbers to adding decimal numbers.
2. Compare subtracting whole numbers to subtracting decimal numbers.
3. Meet with a classmate to discuss effective strategies to help you remember place-value positions when adding and subtracting decimals.

Practice

Arrange these numbers in columns in the correct place-value positions. Estimate each sum. Then add the numbers.

1. 15.42 + 5773.71 + 1.00   2. 577.02 + 71.53 + 0.11

Arrange these numbers in the correct place-value positions. Estimate each difference. Then subtract.

3. 101.43 − 5.37   4. 12.44 − 0.07
5. 11.29 − 0.42   6. 0.88 − 0.08

7. Write the answers from question 1 in words and in expanded form.

NUMBER SENSE AND NUMERATION

Lesson 8
Multiplying Decimal Numbers

WEATHER REPORT

PLAN:
You will learn to multiply decimal numbers by single digit whole numbers and mentally by 10.

DESCRIPTION:
Many weather stations across Canada collect number information that allows us to see weather patterns and record extreme weather.

Get Started

Kenny knows that 21.95 cm of rain fell in 12 h at Jordan River Diversion, BC, and wants to know how many centimetres of rain fell in 36 h, or 3 times that duration.

Vocabulary

duration: How long something takes or lasts
rate: A relationship between quantities measured in different units. For example, 35 km/h means that something will travel 35 km in one hour.

Multiplying decimals is very much like multiplying whole numbers.

Numbers	Number of Decimal Places
21.95	2 places
× 3	+ 0 places
65.85	2 places

65.85 cm of rain would fall in 36 h, if the rainfall continued at this rate.

When multiplying decimal numbers, do the following:

a) Multiply decimal numbers as you would whole numbers.
b) Count the total number of places to the right of the decimal point in the two numbers you are multiplying.
c) Use the total from step (b) to count the number of decimal places in your product. Remember to count from right to left.
d) Fill in with 0s if more decimal places are needed.

Chapter 1: Weather Numbers

1. If you were multiplying 21.95 by 0.3 instead of 3, how many decimal places would be in your product? Explain why. What would your answer be?

2. Apply what you know to multiply these numbers.

   a) 0.4   b) 0.56   c) 93.71
      × 9   × 7   × 6

3. Estimate your answer for each question in 2. Are your answers reasonable? Explain why or why not. Check your estimate and calculations with a calculator.

Tip
Review the steps you followed when multiplying two whole numbers. How can they help you multiply decimals?

Build Your Understanding

Multiply Decimal Numbers

1. Use what you know about basic multiplication facts to answer the following:

   a) 7 × 0.7   b) 0.7 × 6   c) 11 × 1.1   d) 1.1 × 4
   e) 0.5 × 5   f) 0.5 × 9   g) 10 × 1.2   h) 1.2 × 10

2. a) How are whole-number basic multiplication facts similar to decimal basic multiplication facts? How are they different?

   b) What strategies did you use to remember where to place the decimal point correctly? Meet with a classmate to compare and discuss your results.

3. Find the products.

   a)   7   b)   7   c)   9   d)   9
      × 0.1   × 0.01   × 0.1   × 0.01

4. Make note of any patterns you notice in your products. Compare and discuss your conclusions with a classmate.

Lesson 8: Multiplying Decimal Numbers

5. Multiply the following:

   a) 0.4 × 8
   b) 51 × 0.7
   c) 39.2 × 6
   d) 182 × 0.4
   e) 90.3 × 5
   f) 77.7 × 9
   g) 84.3 × 7
   h) 39.2 × 6
   i) 49.2 × 4
   j) 19.03 × 5

Tip
Check your final calculations with a calculator.

What Did You Learn?

1. How is multiplying whole numbers similar to multiplying decimal numbers? How is it different?

2. Discuss with a small group of classmates any patterns you noticed that will help you when multiplying decimal numbers.

Practice

1. Create a decimal-number basic multiplication facts chart. For example, instead of including the 11 × 9 facts on the chart, use 1.1 × 9.

Technology

Using a word processing program, make a list of ways that we use decimals in our everyday lives. Group and rearrange the items according to their similarities, using the cut and paste feature. Look for examples at home and school.

Multiply the following:

2. 0.6 × 8
3. 7.3 × 9
4. 139.7 × 4
5. 6.42 × 3
6. 171.85 × 5

Multiply these numbers:

7. 0.17 × 9
8. 9.77 × 6
9. 0.48 × 3
10. 56.3 × 8
11. 342.71 × 7

12. Estimate to check whether each product in questions 7 to 11 is reasonable. Check your final calculations with a calculator.

Chapter 1: Weather Numbers

Lesson 9
Dividing Decimal Numbers

WEATHER REPORT

PLAN:
You will learn to divide decimal numbers and solve problems that involve dividing decimal numbers.

DESCRIPTION:
In India, very large land and sea winds called "monsoons" affect weather. In summer, southwest monsoon winds carry wet air from the sea, which often creates sudden, heavy monsoon rains. In Cherrapunji, India, for example, it once rained 4.8 m in 15 days!

Quthbullapur, India

Get Started

During a very severe monsoon, a community in India received 2.24 m of rain in 7 days. At that rate, about how many metres of rain fell each day of the week?

Dividing decimal numbers is very much like dividing whole numbers.

Dividing Whole Numbers

$$\begin{array}{r} 32 \\ 7\overline{)224} \\ \underline{21} \\ 14 \\ \underline{14} \\ 0 \end{array}$$

Dividing Decimals

$$\begin{array}{r} .32 \leftarrow \text{quotient} \\ \text{divisor} \rightarrow 7\overline{)2.24} \leftarrow \text{dividend} \\ \underline{21} \\ 14 \\ \underline{14} \\ 0 \end{array}$$

When dividing a decimal by a whole number, do the following:

a) Put the decimal point in the quotient, or answer, exactly above the decimal point in the dividend.

b) Divide as you would with whole numbers.

About 0.32 m of rain fell each day during the monsoon.

Meet with a partner and do the following:

1. Round the dividend to the nearest tenth. What basic division fact will help you estimate the answer? Based on your estimate, is the quotient 0.32 reasonable? Explain why or why not.

2. Multiply to check your answer.

3. Double-check your quotient using a calculator.

> **Tip**
> Review the steps you followed when dividing whole numbers by one-digit divisors. Use grid paper when doing calculations to make sure that you place the decimal point in the correct position.

Build Your Understanding

Divide Decimal Numbers

1. Work with a partner. Take turns answering these division questions. Explain to each other how you got your answers for the questions with decimals.

    a) 63 ÷ 7   b) 6.3 ÷ 7   c) 12.1 ÷ 3   d) 14.4 ÷ 2
    e) 13.2 ÷ 4   f) 10.8 ÷ 2   g) 0.63 ÷ 7   h) 1.21 ÷ 1
    i) 1.44 ÷ 2   j) 1.32 ÷ 2   k) 1.08 ÷ 2   l) 27.8 ÷ 2

2. Divide:

    a) 30.5 ÷ 5   b) 6.48 ÷ 8   c) 12.8 ÷ 4
    d) 5.67 ÷ 9   e) 6.66 ÷ 6

3. Use an estimation strategy to decide whether your quotients are reasonable.

4. The record rainfall for a 6-h period on Vancouver Island is 139.4 mm. This amount was recorded at Ucuelet Bynor Mines in 1965. About how much rain fell each hour?

Chapter 1: Weather Numbers

What Did You Learn?

1. How is dividing whole numbers similar to dividing decimal numbers? How is it different?

2. What useful strategies did you use to remember where to put the decimal place when dividing decimal numbers?

3. What are your strengths when solving problems? How could you improve your problem-solving skills?

Practice

1. 7.34 ÷ 2   **2.** 8.49 ÷ 3   **3.** 5.10 ÷ 5   **4.** 42.24 ÷ 4

5. 6.16 ÷ 7   **6.** 15.11 ÷ 2   **7.** 39.81 ÷ 3   **8.** 22.06 ÷ 8

9. 4.09 ÷ 6   **10.** 27.17 ÷ 4   **11.** 19.19 ÷ 9   **12.** 14.07 ÷ 6

Extension

13. 7.34 ÷ 2   **14.** 8.49 ÷ 3   **15.** 5.10 ÷ 5   **16.** 42.24 ÷ 4

Lesson 10

Mental Math Strategies

WEATHER REPORT

PLAN:
You will learn and develop mental computation strategies to solve number problems.

DESCRIPTION:
Heavy rains and melting snow are just some of the causes of floods. In 1997, people who lived in the valley of Manitoba's Red River were affected by a major flood. About 30 000 people had to leave their homes when the Red River rose 7.5 m above its normal level. People living in the area and 8400 Canadian soldiers made more than 6 000 000 (six million) sandbags to protect communities from flood damage.

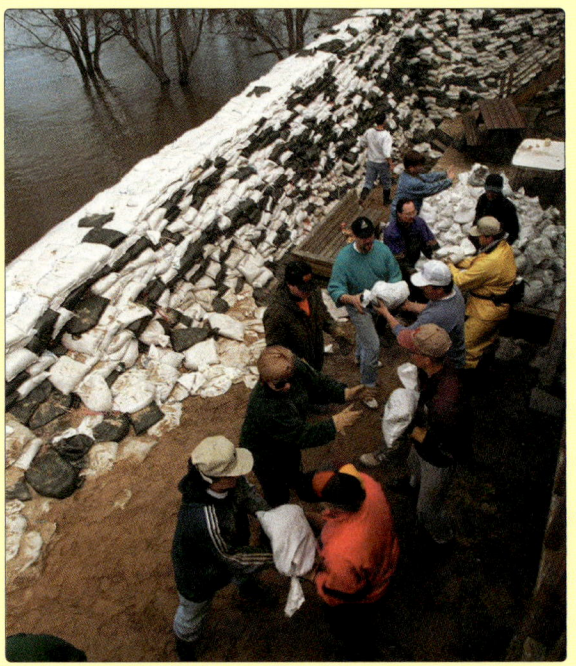

Red River, Manitoba

Get Started

Often you need to do calculations without paper and pencil or a calculator. These calculations are called mental calculations. There are a number of strategies that can help you perform mental calculations quickly and accurately.

The drawing on the right shows how many sandbags were made to protect the community from flood damage. How might you calculate the number of sandbags without counting them? Share and discuss your answers with the class.

I wonder how many sandbags there are.

Chapter 1: Weather Numbers

Clustering

Clustering is a strategy used to estimate the sum of numbers that cluster around a particular amount. For instance, the numbers 22, 27, 26, and 23 cluster around 25, so you can estimate that the sum of the four numbers is about 100.

You are asked to find the product of 2 x 8 x 5. This might be too difficult to calculate in your head. However, if you group the factors (2, 8, and 5) by multiplication facts that you know, it is easier to multiply the numbers mentally:

(2 x 5) x 8

Now you can figure out the answer: 2 x 5 = 10, and 10 x 8 = 80.

Rounding

Rounding is another helpful strategy to use when doing or checking calculations. For example, 107 rounded to the nearest 10 is 110.

1. What is 107 rounded to the nearest 100?
2. What is 96 rounded to the nearest 10? to the nearest 100?

You could use rounding when subtracting numbers.

4871	4871 can round to 4900
− 4104	4104 can round to 4100
	The difference is about 800.

You could also use rounding in division questions.

6392 ÷ 8   6400 ÷ 8

The quotient is about 800.

Tip

Tools and strategies you might use to master basic multiplication and division facts include number lines and skip counting. Share and discuss other useful strategies with classmates.

Vocabulary

clustering: A strategy that can be used to estimate the sum of numbers that cluster around the value of one number. For example, the numbers 44, 45, 54, and 55 cluster around 50. You can estimate that 50 + 50 + 50 + 50 = 200.

estimation strategies: Mental mathematical strategies used to obtain approximate answers

Build Your Understanding

Apply Mental Mathematical Strategies

1. Use clustering to find the sum. Then explain to a classmate how you used the strategy.

    a) 9 + 29 + 11 + 7 + 14 + 13 + 6 = ▇

    b) 17 + 101 + 23 + 18 + 22 + 21 = ▇

    c) 411 + 96 + 103 + 97 + 104 = ▇

2. Create a list of numbers for a classmate to add mentally.

> **Tip**
> If you want to add numbers quickly, look for zeros.
> For example, 60 + 80 = 140.
> 6 + 8 = 14; then add 0.

3. Use the above strategy from the Tip box to answer the following questions.

    a) 70 + 80   b) 120 + 70   c) 220 + 80

    d) 650 + 270   e) 70 + 40   f) 90 + 20

    g) 160 + 120   h) 270 + 230

If you know your basic addition, subtraction, multiplication, and division facts, you can do many other calculations more quickly and accurately. For example, if you know that 4 x 8 is 32, you can easily calculate 4 x 80.

4. Calculate the answer to each question mentally. Beside each answer, write the basic fact that helped you get the answer.

    a) 9 + 51   b) 48 – 30   c) 60 x 4

    d) 120 ÷ 4   e) 9 x 900

Factoring (finding the numbers that multiply together to get a product) can help you calculate mentally. For example, you might find 24 x 25 difficult to do in your head. However, if you think 6 x 4 x 25 or 6 x 100, the question becomes easier to do.

5. Use factoring to multiply these numbers mentally.

    a) 25 x 36   b) 20 x 35   c) 75 x 9   d) 18 x 45

6. Use clustering to solve the following.

    a) 4 x 7 x 5   b) 2 x 8 x 10   c) 4 x 15 x 25   d) 6 x 14 x 5

What Did You Learn?

1. Which strategy or strategies did you find most useful? Why?

2. Explain how knowing the rules for multiplication and division by 10 and by 100 can help you calculate mentally.

3. Which mental computation strategy or strategies would you use to answer the following question? Explain why you selected the strategies you did. Then, find the quotient: 4893 ÷ 9

Practice

Calculate the answer to each question mentally. Beside each answer, write the basic fact that helped you to get the answer.

1. 93 + 7   **2.** 68 − 40   **3.** 70 x 4   **4.** 160 ÷ 4

Use factoring to multiply these numbers mentally

5. 15 x 25   **6.** 15 x 45   **7.** 36 x 9   **8.** 24 x 48

Use clustering to answer the following:

9. 5 x 6 x 5   **10.** 2 x 9 x 5   **11.** 4 x 21 x 25   **12.** 18 x 27 x 5

Show What You Know
Review: Lessons 7 to 10, Mental Math and Operations With Decimals

1. Use mental math strategies to solve the following:

    a) 8 x 20   b) 2 x 12 x 25
    c) 75 x 6   d) 2 x 30 x 5

2. Winds called "prevailing westerlies" can cool the air temperature by about 0.8°C for every 100 m they rise. How many degrees will they cool if they rise 400 m? Work with a classmate to develop a problem-solving strategy. What is the important problem information? What operation or operations will you need to use? Why? How might drawing a labelled picture help? What pattern(s) do you notice?

3. Here are the amounts of snow that fell over 5 days: 18.67 cm, 21.47 cm, 18.87 cm, 24.09 cm, and 23.89 cm. Calculate the mean amount of snow that fell per day. Explain how you got your answer.

NUMBER SENSE AND NUMERATION

Lesson 11

Solving Two-Step Number Problems

WEATHER REPORT

PLAN:
You will develop a plan for solving one-step and two-step number problems.

DESCRIPTION:
Sometimes temperatures reach record highs, the highest temperatures recorded in specific places. It is interesting to compare these records among provinces and even countries.

Get Started

When solving a problem, it is important to have a clear problem-solving plan. Here is a plan that will help you reach the solution. A good problem-solving plan includes these important stages:

1. *Understand the Problem*

    Read the problem carefully to answer these questions:

    a) What are you asked to find?

    b) What information are you given?

2. *Make the Plan/Pick a Strategy*

    a) What operations must you use?

    b) What calculations do you need to do? Why?

    c) What strategy can you use to solve the problem?

3. *Solve the Problem*

    a) Do your calculations.

    b) Use pictures, numbers, and words to explain your answer.

4. *Share and Reflect*

    a) Does your solution seem reasonable? Why?

    b) How can you check your solution?

Some problems are one-step problems. With these problems, you will need to use one operation—addition, subtraction, multiplication, or division.

Problem 1 is a one-step problem.

Problem 1: The highest maximum air temperature in the world, 58°C, was recorded in Al'Aiziyan, Libya, on September 13, 1922. The highest maximum air temperature in Canada, 45°C, was recorded in Midale, Saskatchewan, on July 5, 1937. What is the difference between these two record-high temperatures?

In a two-step problem, you must use two operations. Problem 2 is a two-step problem.

Problem 2: What is the average warmest temperature of the three provinces in the table below?

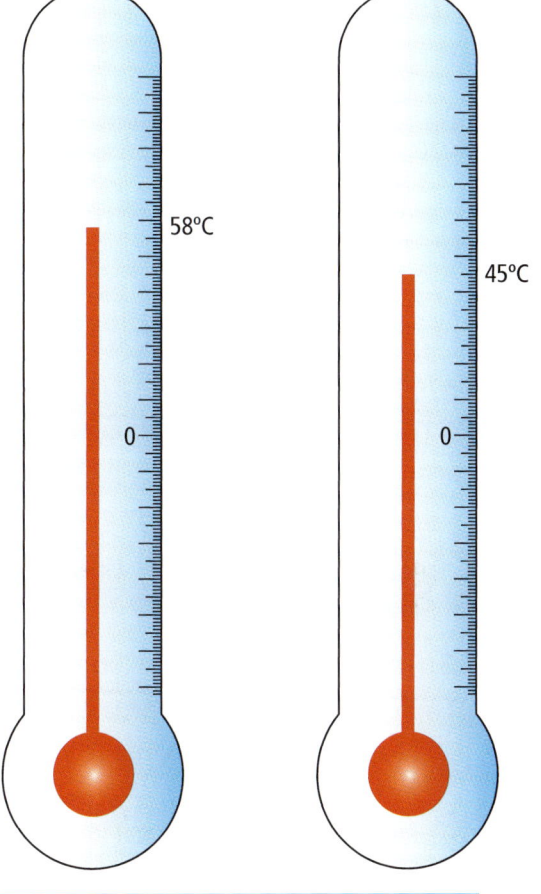

1. What are the two operations you must use when calculating the mean?

2. Work with a classmate. Use the problem-solving plan from Get Started to solve Problems 1 and 2.

Warmest Temperatures in Canada			
Place	Province	°C	Date
Northwest River	Newfoundland	41.7	August 11, 1914
Charlottetown	Prince Edward Island	36.7	August 19, 1935
Nepisiguit Falls	New Brunswick	39.2	August 18, 1935

3. Meet with a small group of classmates to share, compare, and discuss your solutions and problem-solving strategies.

Vocabulary

operation: Addition, subtraction, multiplication, or division
strategy: A plan to help you solve problems

Lesson 11: Solving Two-Step Number Problems

Build Your Understanding

Solve One-Step and Two-Step Problems as a Group

Work as part of a problem-solving team. Follow the problem-solving plan from Get Started to solve each problem. Feel free to improve the plan by adding more steps or asking other questions. Once you arrive at a solution, use estimation to check whether your answer is reasonable.

Tip

Each group member might lead the group through a different stage of problem solving. For example, the first group member might make sure everyone in the group reads the problem carefully and understands what it is asking.

1. The world record for the greatest precipitation in one month is 9300 mm in Cherrapunji, India, in July of 1861. The record in Canada is 2235.5 mm, which fell in Swanson Bay, British Columbia, in November of 1917. How much more rain fell in Cherrapunji than in Swanson Bay during their record months?

2. The greatest annual snowfall in the Yukon was 365.7 mm, in the Northwest Territories it was 234.5 mm, and in Nunavut it was 602.4 mm. How much snow fell in total in these three territories?

Tip

Review and use your chart of "signal" words and phrases that indicate which operations to use in number problems to help you solve these problems.

3. Find the mean high temperature for the four locations on this chart.

Warmest Canadian Record Temperatures			
Place	Province	°C	Date
Collegeville	Nova Scotia	38.2	August 19, 1935
Ville Marie	Québec	40.0	July 6, 1921
Biscotasing	Ontario	42.2	July 20, 1919
St. Albans	Manitoba	44.4	July 11, 1936

4. If you combined the data on this chart with the data on the chart in Get Started, would the mean increase or decrease? Explain your reasoning. Then check your prediction using a calculator. Round your answer to the nearest tenth.

Chapter 1: Weather Numbers

What Did You Learn?

Work with your group to discuss and answer the following questions:

1. Did using mental computation help you solve the problems? Explain.
2. How well did the problem-solving plan work?
3. How would you improve the plan next time?
4. Why would you make these changes?

Practice

Math Problems to Solve

Apply your improved plan to solve these problems.

1. The world's heaviest hailstone fell in Guangdong province, China, in 1995. Its mass was 15 000 g. Canada's heaviest hailstone fell in Cedoux, Saskatchewan, in 1973. Its mass was 290 g.

   a) How long after the Saskatchewan hailstone fell did the Chinese hailstone fall?

   b) How much more was the mass of the Chinese hailstone?

2. On Vancouver Island, the most rain that ever fell in 30 min was 35.6 mm. The most rain that ever fell in one hour was 48.8 mm.

   a) If the 30-min rate continued for an hour, would it be greater or less than the record for one hour?

   b) What would be the difference between the two amounts for an hour?

Lesson 11: Solving Two-Step Number Problems

Chapter Review

1. Give the place value of each of the underlined digits.
   a) <u>5</u>73.69   b) 573.6<u>9</u>   c) 5<u>7</u>3.69
   d) 57<u>3</u>.69   e) 573.<u>6</u>9

2. Order these numbers from least to greatest.
   132.5   132.47   131.96   133.01   133.1

3. Complete the chart in your notebook.

	Standard Form	Expanded Form	Written Form
a)	342 501		
b)		900 000 + 40 000 + 300 + 80 + 6	
c)			four hundred thousand seven hundred eighty-two

4. Write these fractions as decimal numbers:
   a) $\frac{4}{10}$ b) $\frac{4}{100}$ c) $\frac{7}{10}$ d) $\frac{6}{100}$

5. Write these decimals as fractions:
   a) 0.5   b) 0.05   c) 0.15   d) 0.70

6. Change all these numbers to decimal numbers; then order them from greatest to least.
   a) six tenths   b) $\frac{6}{100}$ c) $\frac{66}{100}$ d) sixteen hundredths

7. Complete these patterns:
   a) 11, 22, ... 132   b) 12, 24, ... 144

Chapter 1: Weather Numbers

8. Add or subtract these numbers:

a) 3961
     284
 + 7846

b) 5381
    2974
 + 3462

c) 6473
  − 2964

d) 8342
  − 5999

 9. Multiply or divide these numbers:

a) 82
   x 7

b) 64
  x 35

c) 959 ÷ 7

d) 1496 ÷ 8

 10. Add or subtract these decimal numbers:

a) 147.09 + 13.2 + 2378.46

b) 32.13 + 1.7 + 437.5

c) 1537.61 − 181.9

d) 4376.01 − 2487.09

 11. Multiply:

a) 0.7
   x 5

b) 483.9
    x 4

c) 14.86
    x 9

 12. Divide:

a) 7.2 ÷ 9

b) 781.2 ÷ 2

c) 5 ÷ 0.25

13. Use mental math strategies to solve:

a) 8 x 700

b) 4 x 9 x 25

c) 20 x 45

d) 360 ÷ 6

Chapter Review

NUMBER SENSE AND NUMERATION
DATA MANAGEMENT AND PROBABILITY

Chapter Wrap-Up

Make a Comparison Weather Chart or Poster

You have reached the end of the first chapter in this unit. Throughout the chapter, you learned how mathematics is related to weather. Most importantly, you learned the following mathematics topics and skills: whole and decimal numbers, place value, operations with whole numbers and decimals, and how to develop and apply useful problem-solving strategies.

Your task now is to work as part of a class weather-station team to complete a chart or poster comparing weather information throughout the world, across Canada, and in your school community. Then you will create and solve problems relating to the chart or poster.

Charts are great tools to use when you need to organize and compare information, such as weather records. You might wish to model your chart on the unfinished chart shown below.

Comparing Weather Records			
Weather Record	World	Canada	Local

Temperature
   High
   Low
	58°C
	– 89°C		

Precipitation
   Rain (in one year)
   Snow
	26 461.2 mm
	3110.2 cm
		8122.4 mm
		2446.5 cm	

Wind Speed (km/h)	371 km/h	201.1 km/h	
Hours of Sunlight			

1. Research any world and Canadian records missing from the above table. Check all records given in the chart, since records may have been recently broken.

Chapter 1: Weather Numbers

2. To gather local weather information for your chart, set up a class weather station. You will need to find or make weather-recording tools. For example, you might *find* a thermometer and *make* a simple wind-speed gauge.

Tip

Divide researching, constructing, and recording tasks equally among group members.

3. Set a period of time during which you will record weather information. Record local weather information during that time.

4. Review your weather readings. Enter weather records, such as temperature highs and lows, in the correct chart cells.

5. Use the information on the chart to make challenging questions and problems. Include whole and decimal numbers in your questions and problems. Create some problems that require more than one step to solve.

6. Challenge classmates to solve the problems you created and to explain their problem-solving strategies.

7. Decorate your completed chart or poster with examples of extreme weather events you learned about in this chapter, such as lightning, tornadoes, floods, and monsoons.

8. Share, compare, and discuss your completed chart or poster with the class; then display your chart or poster on a class bulletin board.

Technology

Complete the Chapter Wrap-Up using on-line and multimedia resources. Creating your chart or poster with a word processing program or desktop publisher will give your assignment a very professional look.

Chapter 2

Weather Data

Mathematics allows us to construct graphs that tell us about past weather conditions. Mathematics also helps us to predict future weather. Weather, like mathematics, greatly affects the lives and activities of all people.

In this chapter, you will
- convert measurement units using metric prefixes
- read and write time according to 12-h clocks
- collect, organize, and analyze data
- estimate, count, and write amounts up to $1000, and make change for amounts up to $100
- collect, organize, and analyze data
- read and write dates using SI notation

At the end of this chapter, you will use the math skills you have learned to make a graph of an important weather pattern.

Here are some ways you can start thinking about how weather affects us.

1. Think about a time when weather spoiled something you had planned. Share your thoughts with a partner.

2. Think about a time when weather allowed you to do something fun. Share your thoughts with a partner.

Lesson 1
Converting Metric Measurements

WEATHER REPORT

PLAN:
You will learn about metric prefixes and make conversions between metric units based on your knowledge of these prefixes.

DESCRIPTION:
You will often hear or read about weather described in different measurement units. Here are some examples:
- The least amount of precipitation in a year was 12.7 mm (millimetres), which was recorded at Arctic Bay, Northwest Territories.
- During the 1971/72 winter season, 2446.5 cm (centimetres) of snow fell in Revelstoke, British Columbia.
- In warmer climates, high summer clouds are often found at 13 000 m (metres) or 13 km (kilometres).

Get Started

Some basic units of the metric system are the metre, which is used to measure distance, the gram, which is used to measure mass, and the litre, which is used to measure the volume of a liquid.

Other metric units are formed by adding prefixes to these basic units. For example, kilometres, centimetres, and millimetres are often used in weather measurements. The chart on the next page gives the meanings of common metric prefixes and shows how they are related.

Prefixes in Metric Measurement			
Prefix	Linear Measurement	Relationship to a Metre	Abbreviation
kilo-	kilometre	1000 times	km
deca-	decametre	10 times	dam
deci-	decimetre	$\frac{1}{10}$ times	dm
centi-	centimetre	$\frac{1}{100}$ times	cm
milli-	millimetre	$\frac{1}{1000}$ times	mm

Meet with a small group of classmates to discuss these questions.

1. How are the metric measurement units related?
2. How does adding each prefix change the meaning of "metre"?
3. Share your conclusions with the class.

Journal

Write definitions for metric prefixes. For each prefix, give an example of something with that measurement. For example, a fingernail is about a millimetre thick and a centimetre wide. Use a ruler, or other measurement devices, to help you check any examples.

Build Your Understanding

Use Metric Prefixes to Convert Metric Measurements

1. Work in a small group. Make a chart like the one on the next page in your notebook. Make conversion calculations to complete each chart cell.

Metric Conversion			
Millimetres	Centimetres	Metres	Kilometres
1500			
	900		
		60	
			2.5

Chapter 2: Weather Data

2. When your chart is complete, meet with another small group to share and compare your results.

3. Take turns explaining how you made each conversion calculation.

4. Clouds are often classified by how high they are above the earth. The following chart shows at what heights clouds are found during different seasons. In your notebook, make and complete the chart below. Share and compare your results with a partner.

> **Vocabulary**
>
> **centi:** A prefix meaning $\frac{1}{100}$, as in centimetre
>
> **deci:** A prefix meaning $\frac{1}{10}$, as in decimetre
>
> **kilo:** A prefix meaning 1000, as in kilometre
>
> **milli:** A prefix meaning $\frac{1}{1000}$, as in millimetre
>
> **prefix:** A group of letters added to the front of a word that changes its meaning

Cloud Types			
Classification	Season	Metres	Kilometres
high	summer winter	13 000 5 000	
medium	summer winter		7 2
low	summer winter	2000 surface (0)	surface (0)

5. Kenny located information on the greatest average annual precipitation in various provinces of Canada. He found that the record in Nova Scotia was 1630.7 mm, the record in Newfoundland and Labrador was 166.97 cm, and the record in Prince Edward Island was 1.169 m. Which of these provinces had the greatest average amount of rainfall? Order the records from greatest to least. Explain how you got your answer.

Lesson 1: Converting Metric Measurements

6. Use what you have learned about metric prefixes and measurement to estimate weather events in your community, such as a record snowfall or an average annual rainfall. Compare and discuss your estimates with classmates. If possible, check the estimates by researching accurate weather information.

What Did You Learn?

1. Explain how kilometres, centimetres, and millimetres are related to metres and to one another.

2. Explain how you would convert a precipitation reading of 59.05 cm to millimetres. How could you calculate it mentally?

3. How would you convert 59.05 cm to metres? Calculate the conversion.

4. Which metric unit is the best one to use when communicating this precipitation information? Explain why.

5. What other words have you heard with the prefixes "kilo," "centi," and "milli"? Where have you heard or seen these prefixes used, and how were they used?

Practice

Copy each of these words into your notebook. Underline the prefix and explain what it means.

1. milligram
2. centimetre
3. kilogram
4. millimetre
5. kilometre
6. decimetre

Copy and complete the following:

7. 4 m = ■ cm
8. 8971 m = ■ km
9. 1.4 mm = ■ m
10. 1942 cm = ■ m
11. 87 mm = ■ cm
12. 4.2 km = ■ mm

13. Collect rain in a simple rain gauge over a week. Estimate the rainfall you received. Check your estimate by recording the actual rainfall.

Lesson 2
Reading Analog Clocks

WEATHER REPORT

PLAN:
You will practise reading and writing time using analog clocks. As well, you will estimate time intervals to the nearest second.

DESCRIPTION:
Weather readings are collected all over the world, and this information is communicated to many different places. The time when each reading is recorded is important, because this information helps meteorologists see weather patterns forming.

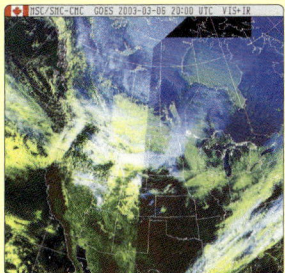 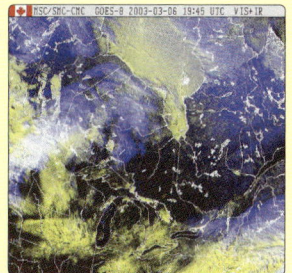
 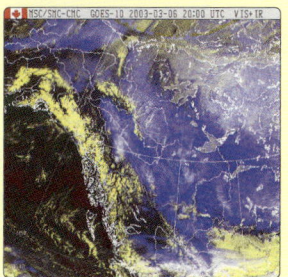

Get Started

An analog clock uses hands to show the time.

You have probably used analog clocks like the ones below to tell time.

Work with a partner. What is the time shown on the left? What is the time on the right?

A.M.   P.M.

Build Your Understanding

Construct an Analog Clock and Read and Write Times

You Will Need
- analog clock picture
- scissors
- glue
- fastener
- construction paper

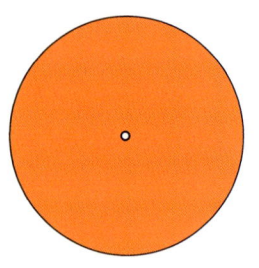 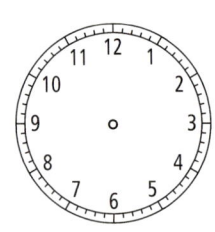 

1. Work with a classmate. Follow the instructions on the analog clock picture your teacher gives you. Cut out the clock parts, and construct an analog clock. Glue the circular parts to construction paper, which will give your clock a firm backing.

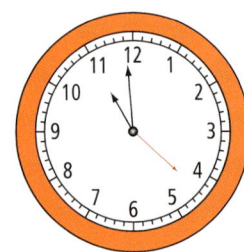

2. Use the fastener to attach the hour, minute, and second hands to the clock face.

3. With your partner, show the following times on the clock.

   a) 4:43   b) 6:18   c) 11:24
   d) 9:16   e) 12:00 f) 12:01

4. Take turns showing times on the clock. Use the second hand to create times to the nearest second. When it is your turn to use the clock, write three different times. Show these times on the model clock to your partner. Challenge him or her to correctly read and then write each time you display.

Journal

Make a note of any clock times you have difficulty reading or writing. Draw these times on sketches of clocks. Record any strategies you might use to help you read, write, and remember these times.

What Did You Learn?

1. What times do you find easiest to read on the clock? Why?
2. What times do you find most difficult to read and write? Why?
3. What are some advantages and disadvantages of using an analog clock with the study of weather?

Practice

Extension

1. You are going to observe the weather throughtout various times in one day. Write down 10 different times throughout the day when you will make observations about the weather. Look outside at those times and record the weather conditions. For example, if one of the times you pick is 8:30 A.M., you might record the weather conditions as "cloudy and cool." Record your observations in a chart. Share, compare, and discuss your chart with a partner.

Estimating Tornado Time

You Will Need
- stopwatch, or a clock or a wristwatch with a second hand

2. Work in a small group. Role-play a situation in which you are in a storm cellar, waiting for a tornado to pass.

   a) One group member takes the role of the tornado and operates the watch. He or she says when the tornado starts, and then when it stops, and records how long the tornado lasted to the nearest second.

   b) Students in the "storm cellar" estimate how long each tornado lasted to the nearest second.

   c) Change roles and dramatize tornadoes that last different time intervals.

   d) Comment in your journal about how well you estimated time intervals. Did your estimates improve after studying a number of tornadoes? Explain why or why not.

Tip

Make the role playing as realistic as possible by researching how long actual Canadian or world tornadoes lasted.

Technology

View or print out a map from the Internet that shows the various time zones around the world. Calculate the current time in different locations around the world. Share and compare your answers with a partner.

Lesson 3
Using SI Notation

WEATHER REPORT

PLAN:
You will learn to write dates according to SI notation, an international standard for communicating information.

DESCRIPTIPTION:
On July 31, 1987, a tornado hit Edmonton, Alberta, causing more than $250 million in property damage. People who study weather record information about weather events and make note of the exact dates when these events happened.

Get Started

SI notation is an international system for writing measurements, including times and dates. SI is the abbreviation for the French term Système International d'Unités. You also use SI when measuring centimetres or kilograms.

There are clear rules for writing dates using SI.
- Write the four digits of the year.
- Leave a space and write the number of the month; if there is only one digit in this number, place a 0 before it.
- Write the number of the day; if there is only one digit in this number, place a 0 before it, too.

According to these rules, the date of the Edmonton tornado is written as follows:

1987 07 31

1. Write your own birthday according to SI notation.

2. Write today's date in SI notation.

Chapter 2: Weather Data

3. On April 20, 1996, two tornadoes struck areas north and south of Shelburne, Ontario. Write this fact using SI notation for the date.

4. Compare and discuss with a classmate the dates you wrote. Check that all dates are written correctly.

Build Your Understanding

Create a Timeline of Important Weather Events Using SI Notation

You Will Need
- long piece of paper for making a timeline
- books on Canadian and world weather

Choose a weather topic that interests you, such as blizzards, floods, hurricanes, etc. Then, do the following:

1. Research the dates of at least five important facts related to your topic. Write the dates using SI notation. For example, if you were interested in tornadoes, one of your facts might be the following:

1912 06 30: The Regina "Cyclone" caused about $4 million in damage.

2. Include at least two Canadian weather facts among the five you collect.

3. When you have collected all needed weather facts, organize the dates from earliest to most recent to create a weather timeline.

4. Meet with a small group to compare, share, and discuss your weather timelines.

Technology

Complete the Build Your Understanding question using a timeline software application or desktop publishing program. Store your data. This way, you can easily add new facts or revise the ones you already have.

Lesson 3: Using SI Notation

What Did You Learn?

1. Write June 22, 1922, in SI notation.
2. Where have you seen or used SI notation dates?
3. Why do you think it is important that people use SI notation when writing dates?
4. Why might it be especially important for people who study weather?

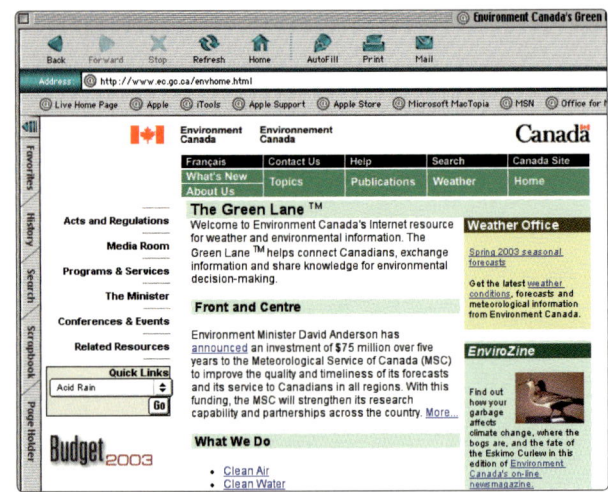

Practice

Write the dates of the following weather facts using SI notation.

1. In Elgin, Manitoba, 52 geese were killed by a thunderstorm on April 22, 1932.
2. On February 16, 1959, a snowstorm in Newfoundland left 70 000 people without services and blocked roads with five-metre drifts.
3. Summerside, Prince Edward Island, had 43 mm of rain in one hour on August 9, 1980.
4. On January 23, 1947, Smith River, British Columbia, had a record cold temperature of −58.9°C.
5. Write in SI notation three dates that are important to you, such as the birthday of a friend or close relative, or the date you accomplished something you are especially proud of.

Show What You Know

Review: Lessons 1 to 3, Measurement and Time

1. Copy and complete the following:
   a) 7 m = ■ cm
   b) 9999 m = ■ km
   c) 7.1 mm = ■ cm
   d) 1826 cm = ■ m
   e) 52 mm = ■ cm
   f) 1.1 km = ■ mm

2. Write the time shown on each clock below.

   a)    b)

   c)    d)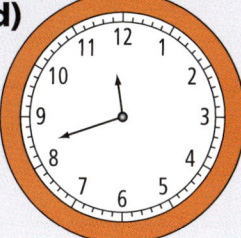

3. Sketch six clocks that show these times:
   a) 9:13
   b) 10:18
   c) 4:05
   d) 8:35
   e) 12:05
   f) 12:43

4. Use SI notation to write down your birthday and the birthdays of all your family members. Organize the dates from oldest to most recent to create a family birthday timeline.

Lesson 4
Exploring Money

WEATHER REPORT

PLAN:
You will learn to estimate, count, and write amounts up to $1000.00. You will also practise making and counting out change for purchases up to $100.00.

DESCRIPTION:
Weather can greatly affect the amount of money people pay to heat and cool their homes.

Get Started

In the illustrations above, you can see the amount paid for gas during different seasons. The total gas bill for the year was $1995.10. Meet in a small group to study the pictures carefully. Answer these questions.

1. How is the temperature outside related to the amount paid for gas to heat the home during each month?

2. Use the amounts given in the pictures above to estimate how much might be paid for gas in winter. Estimate how much might be paid in spring.

3. Share and discuss your estimates with other group members. Do your estimates seem reasonable?

4. What could happen that would make your estimate for any month higher or lower?

Chapter 2: Weather Data

Build Your Understanding

Estimate, Count, and Solve Problems With Money

You Will Need
- play money of Canadian coins and bills

1. Write each amount of money shown below with a "$" sign and a decimal place. For example, a penny is $0.01.

2. Estimate the total amount of money shown below. Then count the bills and coins in the picture to find the total value of the money. How close was your estimate to the actual amount?

Journal
Make notes of estimation strategies you find useful.

Tip
You might use play money to show the amounts pictured in your book. Then use the play money to count the total value of all the coins and bills in the pictures.

Lesson 4: Exploring Money

3. Work in a group. Estimate the amount of money shown in each picture below, and record your estimate. Then count the bills and coins in the picture to find the total value of the money.

a)

b)

c)
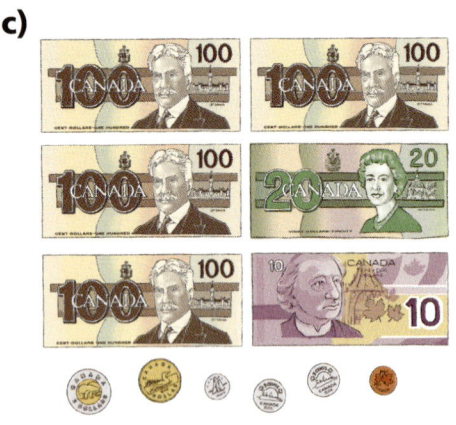

d)
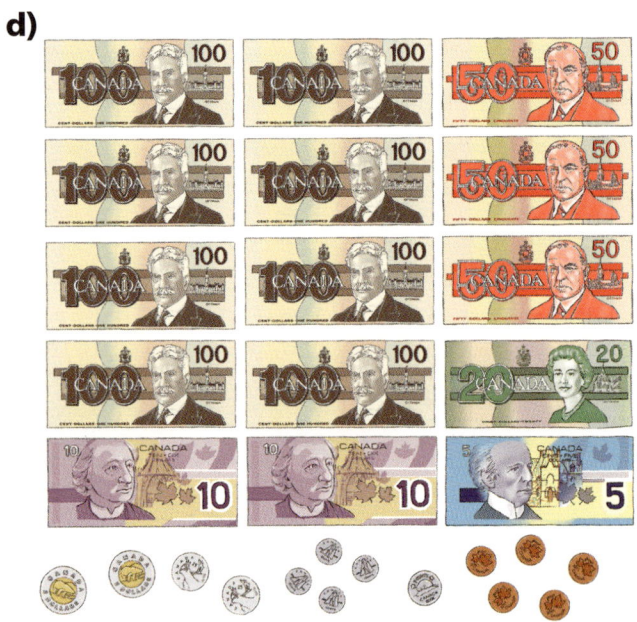

4. Work with a classmate. From your play money, count out $1000.00. Include as many different bills and coins as possible in the total amount. Create an arrangement of money that is less than $1000.00. Challenge your partner to estimate the value of the money and record his or her estimate. He or she then counts the play money and writes the total amount with a "$" sign and a decimal place.

Chapter 2: Weather Data

5. Work with a partner. Use $100.00 of play money in different bills and coins to role-play the following purchasing situations. In one situation, you play the buyer and your partner plays the sales clerk. In the next situation, switch roles.

   Sale Today!

   Sunscreen $4.97
   Earmuffs $9.85
   Waterproof Boots $54.90
   Ski Jackets $97.22

   We Pay All Tax!

   a) You are buying the sunscreen. What is the smallest bill you can use to make the purchase? What change should the sales clerk give you?

   b) You buy the earmuffs and give the sales clerk a $20.00 bill. How much change will you get back?

   c) How many $20.00 bills do you need to buy the waterproof boots? What change will you get back? What other combinations of bills can you use to make the purchase? What change will you get back in each case?

   d) With the different bills and coins you counted out to make $100.00, hand the sales clerk an amount of money as close as possible to the purchase price of the ski jacket. Will you get any change back? If so, how much?

Technology

Use a draw application to show how our coins relate to the metric system of measurement. For example, 1 quarter is 0.25 of 1 dollar.

Journal

Did your ability to estimate money amounts improve as you made more estimates? Why do you think this happened?

What Did You Learn?

1. How is working with money similar to working with decimals?

2. How is it different? Give some examples to support your conclusions.

3. What are your strengths in solving money problems? In what area do you need to improve?

4. What estimating, counting, and calculating skills do you need to work as a sales clerk in a store?

Practice

Math Problems to Solve

1. In Get Started, what was the difference between the most expensive month for gas in the picture and the least expensive month for gas?

2. Tavia was given $75.00 from relatives for her birthday. She wants to buy sunscreen, earmuffs, and waterproof boots, from the sale shown on the previous page, and a $24.50 watch. Estimate then calculate whether Tavia can afford to buy all the items she wants with the money she has. If not, list a combination of items she can buy with the $75.00.

3. Research the prices of special weather clothing or protection items such as waterproof and insulated coats or sunscreen. Make up money problems based on the prices. Your problems must involve estimating, counting, and calculating. Challenge classmates to solve your money problems.

Lesson 5
Data Collection and Analysis

WEATHER REPORT

PLAN:
You will collect weather data about wind direction, organize your data in a tally/frequency chart, represent the findings in a compass-rose graph, and analyze your graph data.

DESCRIPTION:
Environment Canada uses weather vanes to measure wind direction. The equipment is placed on a 10-m-high tower. The tower is built on level ground, well away from anything (such as trees and buildings) that might affect a wind reading.

Electronic monitoring stations will record a wide range of weather data.

Get Started

People who study weather often collect data, organize their data in charts, display the data in graphs, and analyze the graph information.

Ms. Singh's class collected data about September temperatures over a five-day period. They obtained the data shown below.

Monday	Tuesday	Wednesday	Thursday	Friday
15°C	18°C	20°C	21°C	23°C

Lesson 5: Data Collection and Analysis

Ms. Singh's class organized the data using a tally/frequency chart.

Five-Day Temperature Chart		
Temperature	Tally	Frequency
days below 20°C	//	2
days at 20°C or above	///	3

A tally chart uses tally marks to count how many times something happened. The class found that the temperature was at 20°C or above three times, so they made three tally marks on the chart.

Frequency is the number of times something happens. The class counted their tally marks and discovered that the frequency was 3.

Meet in a small group and discuss these questions about the tally/frequency chart.

> **Vocabulary**
> **frequency:** The number of times an event or item occurs
> **graph:** A visual way to represent, or display, data
> **tally chart:** A chart that uses tally marks to count data and record how frequently something happened

1. What is the frequency of temperatures below 20°C?
2. How many tally marks show this frequency?
3. What patterns do you notice in the chart data?
4. If Ms. Singh's class wanted to include Saturday's temperature, which was 22°C, how would it change the tally/frequency chart?

Build Your Understanding

Chart and Graph Wind Direction

You Will Need
- weather vane to measure wind direction
- grid paper
- pencil

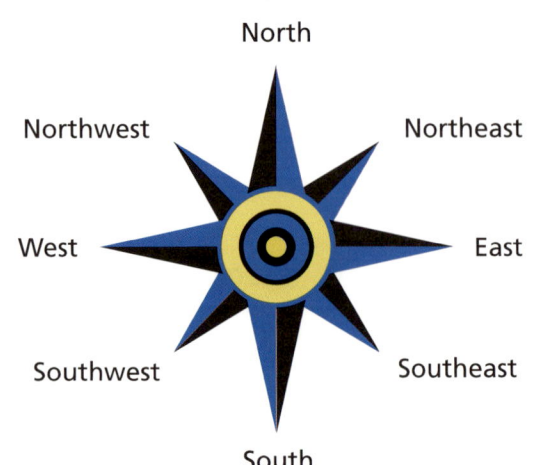

Chapter 2: Weather Data

Create Your Own Weather Vane

You Will Need
- plastic drinking straw
- small piece of cardboard (20 cm x 20 cm)
- straight pin
- glue
- plastic bead with a hollow centre
- pencil with an eraser
- marker
- compass
- masking tape

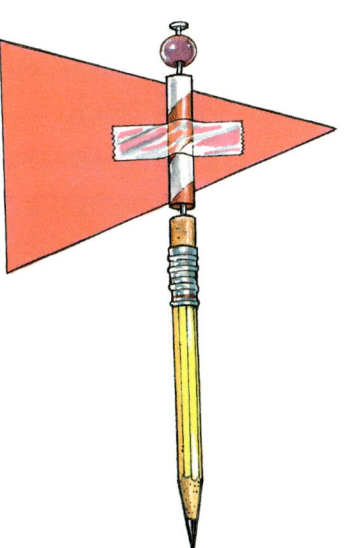

1. In your notebook, make a chart like this one.

Wind Direction		
Wind Direction	Tally	Frequency
north		
northeast		
east		
southeast		
south		
southwest		
west		
northwest		

2. Work in a group or as a whole class. Place the weather vane you have constructed in a good position to collect wind direction data.

3. Establish an amount of time over which you will collect data about wind direction. Two weeks is a reasonable amount of time.

> **Tip**
> Use information about Environment Canada weather vanes to help you position the class weather vane in the best location.

Lesson 5: Data Collection and Analysis

4. Each day during the time period, record the direction of the wind in the appropriate cell of your tally/frequency chart. For example, if the wind comes from the west on one day, place one tally mark on the chart.

5. At the end of the time period, count to find the frequency of wind from each direction.

6. In the middle of the grid paper, place the directions North, Northeast, East, Southeast, South, Southwest, West, and Northwest on an octagon as shown on the right.

7. From the proper side of the octagon, draw one square for each tally mark, moving in the direction of the wind. For example, if you found the wind blew from the north five times, draw five squares in a northerly direction, as in the example shown on the right.

8. Graph all the frequencies on your tally/frequency chart to complete the compass-rose graph.

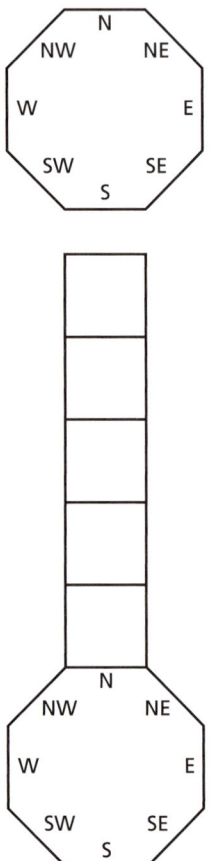

Journal

Make notes of where you have seen weather vanes or windsocks. Why do you think they are used in these locations?

What Did You Learn?

Meet in a small group to discuss these questions.

1. From which direction(s) did the wind blow most often? How can you tell?

2. From which direction(s) did the wind blow least often? How can you tell?

3. What patterns do you notice in your data?

4. What conclusions can you make about the winds in your school area? Why?

Journal

Make notes about how you might improve your work collecting, organizing, graphing, or analyzing data.

Practice

Extension

Predict what your results would be if you collected data about wind direction at another time of the year. Record your prediction. Repeat the data collection and analysis process at that time of the year. Compare your compass-rose graphs at both times of the year. How did your results, as shown in your second compass-rose graph, compare with your prediction?

Technology

Two phenomena that change how the temperature "feels" outside are wind chill and humidex. Using on-line resources, gather information about what these terms mean. Select one of these topics and track the regular temperature over a two-week period, as well as the temperature with the wind chill or humidex. Input both sets of data onto a spreadsheet, and create a line graph to display the information.

Lesson 6

Line Graphs

WEATHER REPORT

PLAN:
You will learn to read and construct line graphs.

DESCRIPTION:
A "chinook" is a dry, warm, strong wind that blows down the eastern slopes of the Rocky Mountains in North America. It can greatly change temperature. For example, on January 6, 1966, at Pincher Creek, Alberta, a chinook raised the temperature 21°C in only four minutes. Meteorologists often collect data on weather and climate change, and show these changes on graphs.

10:35 A.M, temperature −5°C

1:05 P.M., temperature 14°C

Get Started

Ms. Singh's class wanted to show the changes they observed in temperature outside their classroom over a week. They created a graph from this table of temperature data.

Temperatures During the Week of September 9 to 13, 2002		
Time We Took the Temperature		
Date	Time	Temperature
2002 09 09	1:25 P.M.	15°C
2002 09 10	1:25 P.M	18°C
2002 09 11	1:25 P.M	20°C
2002 09 12	1:25 P.M	21°C
2002 09 13	1:25 P.M	23°C

Chapter 2: Weather Data

The class decided a broken-line graph was the best way to present data. When creating the broken-line graph, they drew two axes on grid paper.

The class labelled the horizontal axis "Days." They marked the five days along this axis, using two squares for each day. Just below the horizontal axis, they wrote the names of the five days. Then they labelled the vertical axis "Temperature" and showed temperatures in degrees Celsius from 0 to 26. The students then took turns plotting points to show the temperature on each day. Finally, they drew lines connecting the points together.

Journal

In your own words, write a definition of a "broken-line graph." Include an example of a broken-line graph to support your definition. The graph might communicate your growth in height over months or years. Give the graph a title. In a caption, identify the intervals. Make notes on why you selected these intervals.

Meet in a group or as a class to discuss these questions about the broken-line graph.

1. Why did Ms. Singh's class decide to stop numbering on the vertical axis at 26°C?

2. What pattern or patterns do you notice in the graph data?

3. What can you conclude about the temperature outside Ms. Singh's classroom for this week? Why?

Vocabulary

axis: One of the intersecting number lines on a graph. Two or more of these lines are called axes.

broken-line graph: A graph that displays data by showing points joined by lines

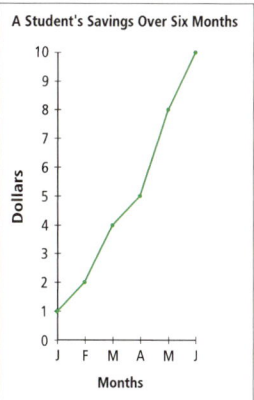

interval: A space between two points. For example, the interval between two points on the horizontal axis in the graph above is one month; the interval between two points on the vertical axis is one dollar.

Lesson 6: Line Graphs

Build Your Understanding

Construct a Broken-Line Graph Showing Temperature Over a Five-Day Period

You Will Need
- thermometer
- grid paper
- ruler
- pencil

Tip
Divide data collection duties equally among class members.

1. Work together as a class to collect temperature readings at the same time each day over a five-day period.

2. Once you have collected your temperature data, organize it into a chart like the one in Get Started.

3. Work with a partner to create a broken-line graph with your data.

   a) Draw horizontal and vertical axes on a piece of grid paper.

   b) Label the horizontal axis and mark off appropriate time intervals.

   c) Label the vertical axis and mark off appropriate temperature intervals.

   d) Plot points using data from your table.

   e) Use a ruler and pencil to connect all the points.

   f) Give your broken-line graph an appropriate title.

4. Meet with other pairs to share, compare, and discuss your graphs.

What Did You Learn?

1. Why did you select the intervals you did along the horizontal and vertical axes?

2. What do you feel is particularly good about your graph? What needs to be improved? Ask a classmate how you could improve your graph.

3. What patterns do you see in your graph data?

4. From your broken-line graph, what conclusions can you draw about temperature during this period? Why?

Practice

Extension

Create a media report in which you communicate the important findings of your temperature study. It could be a short presentation for a television weather channel, a radio broadcast, or a layout for a Web page. Use your broken-line graph as the focus of the media presentation. Make sure that all graph labelling is complete, clear, and accurate.

Vocabulary

survey: A tool for collecting information by asking people questions or interviewing them

Show What You Know

Review: Lessons 4 to 6, Money and Data Management

1. Pretend you are given $100.00 to spend on five items. Use newspapers, flyers, or the Internet to research and record the items you would buy. Estimate then calculate whether you can afford to buy all five items you want with the money you have. If not, list a combination of items you could buy with the $100.00. How much change would you get back?

2. Survey at least 20 students to find out which season they like best. Make a tally/frequency chart to record your data. What conclusions can you make from your data?

3. Find data about weather that you could use to make a good broken-line graph. Construct the graph. Then share and discuss your completed graph with a classmate. Explain important decisions you made as you constructed the broken-line graph.

Lesson 7

Bar Graphs

WEATHER REPORT

PLAN:
You will learn how to read and create bar graphs.

DESCRIPTION:
Bar graphs can give us important weather information, such as when the greatest number of tornadoes strike, when a community has the most fog, and which are the sunniest months of the year.

Get Started

Meet in a small group and carefully read the weather bar graphs shown on the next page.

1. In what month is a tornado most likely to strike in Manitoba?
2. When would you visit Vancouver if you wanted to have the least chance of fog?
3. What are the sunniest three months in Watson Lake?
4. Make up other questions about the bar graphs. Challenge classmates to answer your questions.

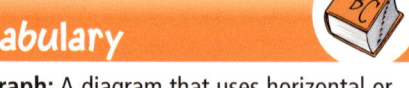

Vocabulary

bar graph: A diagram that uses horizontal or vertical bars to show data

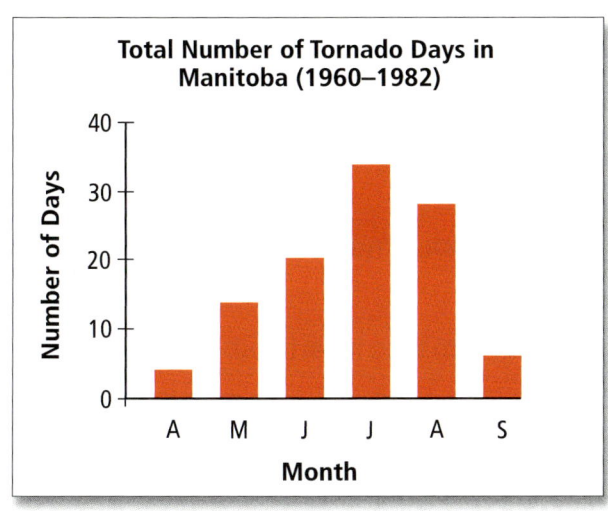

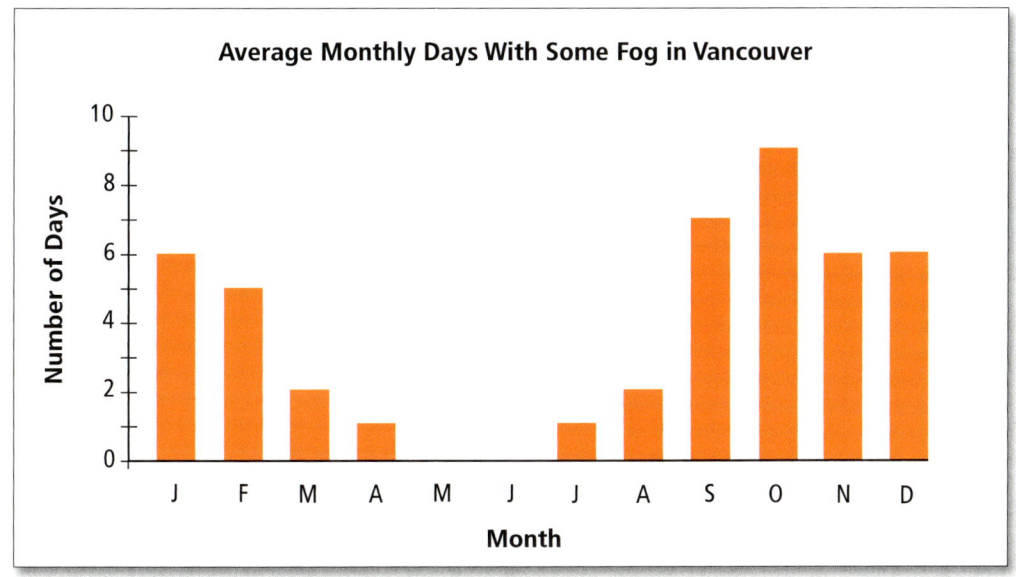

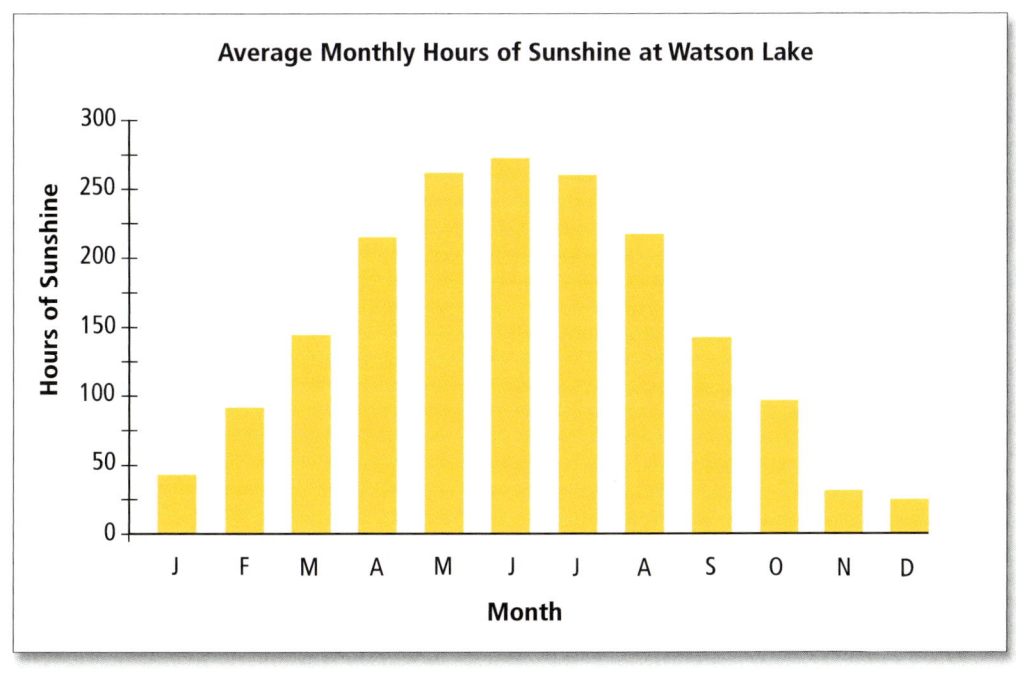

Lesson 7: Bar Graphs 95

Build Your Understanding

Construct a Bar Graph

You Will Need
- grid paper
- coloured pencils

1. Use the data in the chart on the right to make a bar graph of average monthly mid-afternoon temperatures in Vancouver, BC.

   a) Draw horizontal and vertical axes on a piece of grid paper.

   b) Label the horizontal axis and mark off logical month intervals.

   c) Label the vertical axis and mark off logical temperature intervals.

   d) Draw and colour bars to show the average mid-afternoon temperature each month.

   e) Give your bar graph an appropriate title.

   f) Write a paragraph summarizing the most important conclusions you can draw from your bar graph.

2. Meet with a small group of classmates to share, compare, and discuss your bar graphs, and present your paragraphs.

Average Monthly Mid-Afternoon Temperatures in Vancouver, British Columbia	
Month	Temperature
January	5.2°C
February	7.4°C
March	9.4°C
April	12.8°C
May	16.5°C
June	19.2°C
July	21.9°C
August	21.5°C
September	18.3°C
October	13.6°C
November	9.0°C
December	6.5°C

Source: Environment Canada

What Did You Learn?

1. Does the data you used for your graph seem reasonable? Explain why or why not.
2. How did you choose the horizontal intervals on your bar graph?
3. How did you choose the vertical intervals on your bar graph?
4. How would you improve your bar graph after sharing and discussing it with classmates? Why?
5. How might information on the chart affect decisions people make about what clothes to wear?

Journal

Make a checklist of important steps to follow when creating bar graphs and important graph features to include. Use a computer to design the checklist.

Practice

Meet in a group to read the bar graph on the right and discuss these questions.

1. What are the intervals on the horizontal axis? Why do you think these intervals were chosen?
2. What are the intervals on the vertical axis? Why do you think these intervals were chosen?
3. What pattern is shown on the bar graph?
4. What conclusion or conclusions can you draw from the pattern?

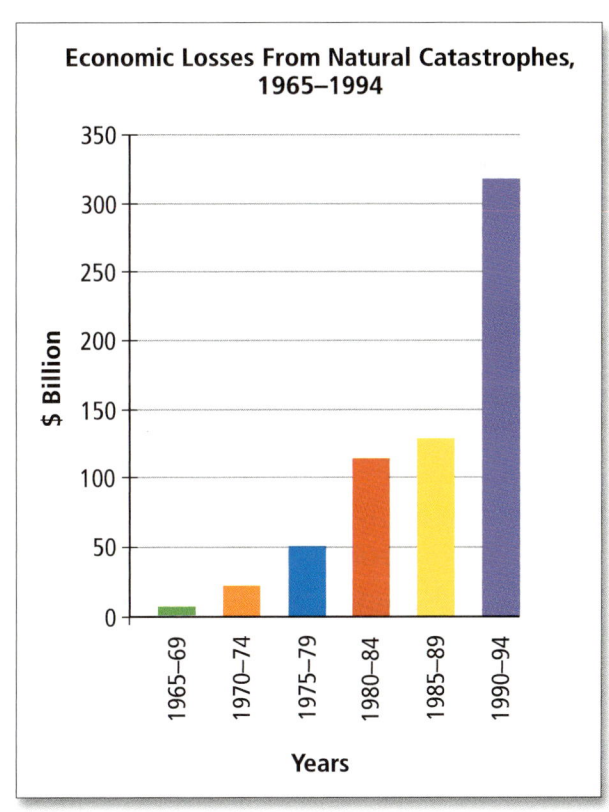

Lesson 7: Bar Graphs

Lesson 8
Pictographs

WEATHER REPORT

PLAN:
You will learn how to read and construct pictographs.

DESCRIPTION:
Hailstorms happen more often in some parts of Canada than in others. These storms can affect many people's lives and do a lot of damage. Often, it is useful to make weather graphs about storms so we can quickly compare information.

Get Started

A good way to organize data, such as information about the number of storms that take place, is in a pictograph. A pictograph is a graph that shows data using picture symbols. Pictographs are useful for making quick comparisons of data.

Average Number of Days per Year With Hail in Western Canada

Winnipeg, Manitoba	○ ○ ○
Estevan, Saskatchewan	○ ○
Edson, Alberta	○ ○ ○ ○ ○ ○
Cape Scott, British Columbia	○ ○ ○ ○ ○ ○ ○ ○ ○ ○ ○ ○ ○ ○ ○

Legend
○ = 1 day with hail

To help your reader clearly understand the pictograph, you must include a legend in which you explain what each picture symbol represents. In the pictograph above, each hailstone represents a day when there was a hailstorm.

Chapter 2: Weather Data

Noma created her weather pictograph using information from this frequency table.

Places in Western Canada Having the Highest Average Number of Days of Hail per Year	
Place	Average Number of Days per Year With Hail
Winnipeg, Manitoba	3
Estevan, Saskatchewan	2
Edson, Alberta	7
Cape Scott, British Columbia	18
Source: Environment Canada	

When creating the pictograph, Noma followed these important steps.
- She decided on a suitable title that described what her pictograph was about.
- She chose a logical symbol (the hailstone) for the information she wanted to present.
- She decided how much each pictograph symbol would represent (1 hailstone symbol = 1 day with a hailstorm).
- She drew each symbol on her pictograph so that all symbols were exactly the same size.

Tip

When reading a pictograph, always do the following:
- first, read the pictograph title
- decide what is being compared
- look at the legend to see what each symbol stands for

Vocabulary

frequency table: A table showing the number of times an event occurs
legend: A feature on a pictograph that explains what each symbol represents, or stands for
pictograph: A graph that shows data using pictures as symbols
symbol: Something, such as a picture, that represents, or stands for, something else

With a classmate, or in a small group, discuss the following questions.

1. Which of the western provinces had the highest average number of days of hail per year? How do you know?

2. Which province had the least? How do you know?

3. How would Noma need to change her pictograph if she decided to let each hailstone symbol represent two days with hail instead of one?

4. In which place would you be most likely to have a hailstorm during your visit? Explain.

Lesson 8: Pictographs

Build Your Understanding

Construct and Read a Pictograph

You Will Need
- grid paper
- coloured pencils

1. Work with a partner. Use the weather data in the frequency table below to create a pictograph.

2. When planning your graph, discuss the pictograph symbol. How could you avoid making your pictograph too cluttered?

3. Meet with other pairs to compare pictographs.

Places in Canada Having the Most Days per Year With Thunderstorms	
Place	Average Number of Days per Year With Thunderstorms
Daniels Harbour, Newfoundland and Labrador	7
Summerside, Prince Edward Island	11
Fredericton, New Brunswick	13
Debert, Nova Scotia	12
St. Hubert, Québec	27
London, Ontario	36
Rivers, Manitoba	26
Wynard, Saskatchewan	25
Edmonton, Alberta	26
Prince George, British Columbia	24

Source: Environment Canada

Chapter 2: Weather Data

What Did You Learn?

1. Explain to a classmate how you used data in the frequency table to create a pictograph.

2. Why did you use the symbol(s) you did? What other symbol(s) might you have used? Why?

3. How is the pictograph like a bar graph? How is it different?

4. What are some advantages of using a pictograph for displaying data rather than a bar graph? What are some disadvantages?

5. What conclusions can you draw about Canadian weather from your pictograph?

Journal

Draw an example of a pictograph about weather. Label all important pictograph features on the graph, such as the title, symbol, and legend. In thought balloons, write reminders that will help you create better pictographs. For example, "Don't use many colours unless they have some mathematical purpose."

Practice

Math Problems to Solve

1. Work in a small group. Identify a problem or question that interests you, such as: What is the most popular winter sport among the students in the class?
   a) Collect data about the problem or question, and record the data on a tally/frequency chart.
   b) Does your data seem reasonable? Explain why or why not.
   c) Create a pictograph from your tally/frequency chart.
   d) Summarize the main conclusion or conclusions you can draw from your pictograph.
   e) Give a presentation about your findings. Use the pictograph as the focus of your presentation.
   f) Discuss ways to improve your pictograph.

2. Create a checklist that will help you to improve your pictographs.

Lesson 8: Pictographs

DATA MANAGEMENT AND PROBABILITY

Lesson 9

Calculating Mean

WEATHER REPORT

PLAN:
You will learn what "mean" is and how to calculate the mean of a set of data.

DESCRIPTION:
In the winter, blowing snow can make it very difficult to travel. In the summer, hot hazy air makes it difficult for some people to breathe.

Get Started

The mean is the average of a set of numbers. To find the mean, first calculate the sum of a set of data. Then, divide by the number of pieces of data.

Anil found the mean high temperature over a five-day period.

He followed the method shown on the next page when calculating the mean.

Temperatures During a Five-Day Period	
Day	High Temperature for the Day
Monday	17°C
Tuesday	16°C
Wednesday	18°C
Thursday	16°C
Friday	18°C

Chapter 2: Weather Data

$$\text{Mean} = \frac{17 + 16 + 18 + 16 + 18}{5}$$

(because there are five temperatures in the set of data)

$$= \frac{85}{5}$$

$$= 17$$

The mean temperature for the five-day period was 17°C.

1. Explain to a classmate how to find the mean temperature for a five-day period with these temperatures: 27°C, 25°C, 26°C, 29°C, and 28°C.

2. Find the mean of the above temperatures.

3. Discuss with your partner how you could use estimation or other mental mathematical strategies to see if your calculations are reasonable.

Vocabulary

mean: The average of a set of numbers. The mean can be found by dividing the sum of all the data by the number of pieces of data.

Journal

"Average" is a synonym for "mean." Synonyms are different words that have the same meaning. Other mathematical synonyms are "multiply" and "times." Make a list of some important mathematical words and their synonyms. Design a chart or graphic organizer to organize the information.

Tip

Try clustering, or grouping, numbers when you add. Check your sums using a calculator.

Lesson 9: Calculating Mean

Build Your Understanding

Calculate the Mean of a Set of Data

1. Use the set of data below to find the mean number of days per year with blowing snow.

Places in Canada Having the Highest Average Number of Days With Blowing Snow per Year	
Place	Average Number of Days per Year With Blowing Snow
Hopedale, Newfoundland and Labrador	45
Greenwood, Nova Scotia	21
Border, Québec	90
Winisk, Ontario	38
Churchill, Manitoba	64
Regina, Saskatchewan	32
Old Glory Mountain, British Columbia	25
Source: Environment Canada	

2. Using the set of data below, find the mean number of days per year with smoke or haze.

Places in Canada Having the Highest Average Number of Days With Smoke or Haze per Year	
Place	Average Number of Days per Year With Smoke or Haze
St. John's, Newfoundland and Labrador	25
Greenwood, Nova Scotia	29
St. Hubert, Québec	118
Windsor, Ontario	228
Winnipeg, Manitoba	30
Saskatoon, Saskatchewan	8
Vancouver, British Columbia	185
Source: Environment Canada	

Chapter 2: Weather Data

3. Work with a partner to find the mean number of days per month with fog using the graph below.

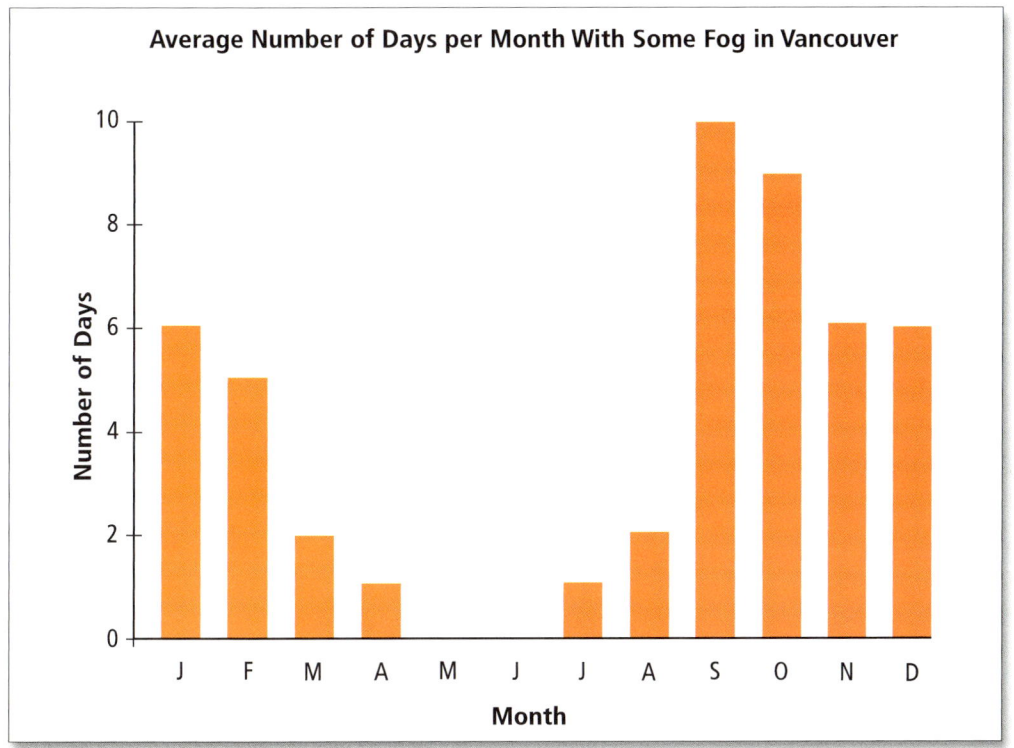

What Did You Learn?

1. What is the mean number in this set of data: 6, 2, 7, 8, 11, 14?

2. What chart data surprised you? Why?

3. How did calculating the mean help you understand the weather data?

Technology

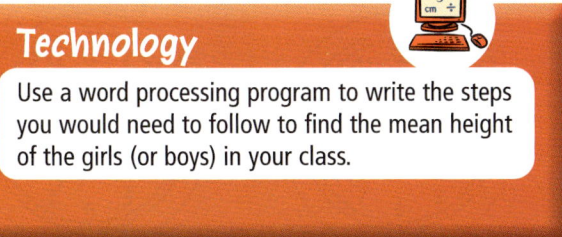

Use a word processing program to write the steps you would need to follow to find the mean height of the girls (or boys) in your class.

Practice

1. Choose one set of data from Build Your Understanding, and use it to create a bar graph. Share your completed bar graph with a classmate, and explain your choice of graph intervals.

2. Use the other set of data from Build Your Understanding to create a pictograph. Share your completed pictograph with a classmate, and explain your choice of graph symbols.

3. Discuss with a group of classmates: Which graph helps you to better understand the weather data—a bar graph or a pictograph? Why?

Show What You Know
Review: Lessons 7 to 9, Data Management

1. Obtain five-day high temperature forecasts for your community from a newspaper, the Internet, or a television or radio broadcast. Use the data to make a bar graph. Write at least two conclusions you can make by looking at your graph.

2. The following tally chart shows how many weather books five students borrowed from the library:

Student	Number of Weather Books
Tahrim	////
Howard	//// //// /
Ming	//// //// //// ////
Brian	/
Joy	//// /

Create a pictograph from the tally chart. Summarize the main conclusion you can draw from your pictograph.

3. Find the mean high temperature of the five-day forecast from question 1.

Chapter 2: Weather Data

DATA MANAGEMENT AND PROBABILITY

Lesson 10

Predicting From Patterns

WEATHER REPORT

PLAN:
You will practise recognizing patterns in weather data tables, and use these patterns to make predictions about weather events.

DESCRIPTION:
Some people believe that if the groundhog Wiarton Willie sees his shadow on February 2, Canadians can expect six more weeks of winter. If he does not, spring is on its way. When making their scientific predictions, meteorologists use the latest technology to gather and analyze weather data.

Get Started

Meteorologists use computers to help them analyze large amounts of weather data and make accurate predictions called "forecasts." The computers can trace patterns in past weather and predict how weather should develop.

The table on the right shows a chart of temperature data for a Canadian community.

Temperature in a Community for Six Days	
Day	Temperature
Monday	22°C
Tuesday	23°C
Wednesday	24°C
Thursday	25°C
Friday	26°C
Saturday	■

Lesson 10: Predicting From Patterns

Work with a partner or in a small group to study the temperature table on the previous page.

1. What pattern do you notice?
2. Predict what the Saturday temperature will be, based on this pattern.
3. What could happen that would make your temperature prediction incorrect?

Vocabulary

prediction: A reasonable guess as to the outcome of an event

Journal

Make notes of any predictions you make in a day. The predictions can be related to weather or other areas of your life. Some possible predictions are "It looks like it's going to rain" and "I bet the Vancouver Canucks will win the game." Underline words or phrases in the sentences you write that indicate that you are making a prediction.

Build Your Understanding

Make Predictions Based on Tables of Weather Data

You Will Need
- charts of Average Monthly Sunshine in Kamloops and Prince Rupert
- coloured pencils

1. Work with a partner. Study the table below and make notes about any patterns you notice.
2. Use what you have learned to predict what the August temperature will be.

Mid-Afternoon Temperature Average in Regina	
April	9.4°C
May	18.2°C
June	22.7°C
July	26.1°C
August	
September	18.6°C
October	11.9°C

Tip

When making your prediction, also draw from your own knowledge about weather at different times of the year.

Chapter 2: Weather Data

3. Study the bar graphs below for monthly sunshine in Kamloops and Prince Rupert. In your notebook, write about any patterns you notice.

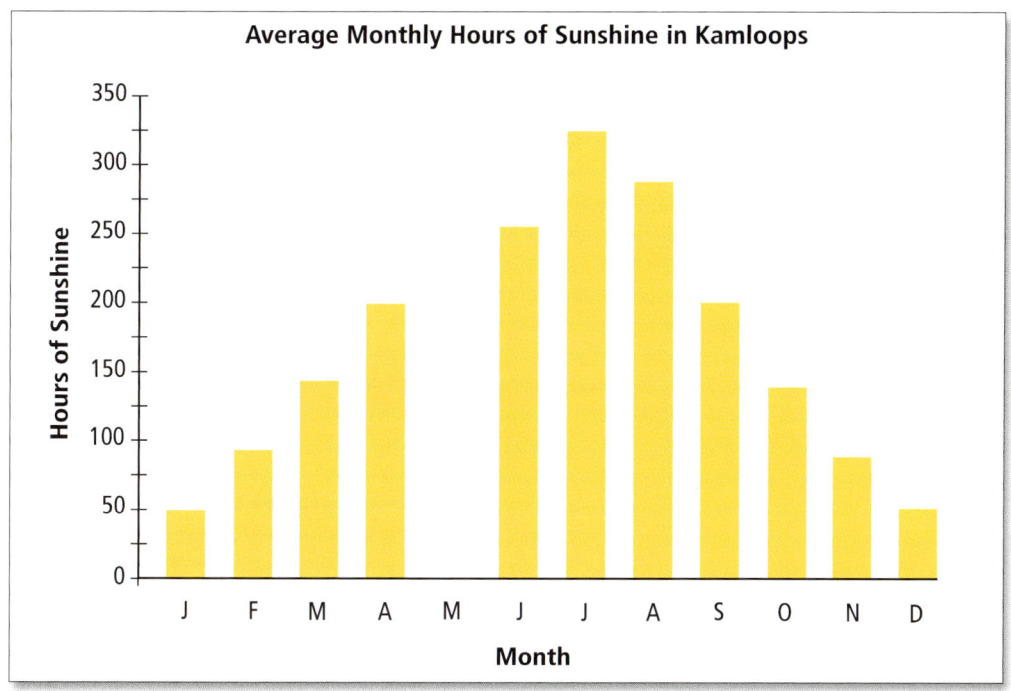

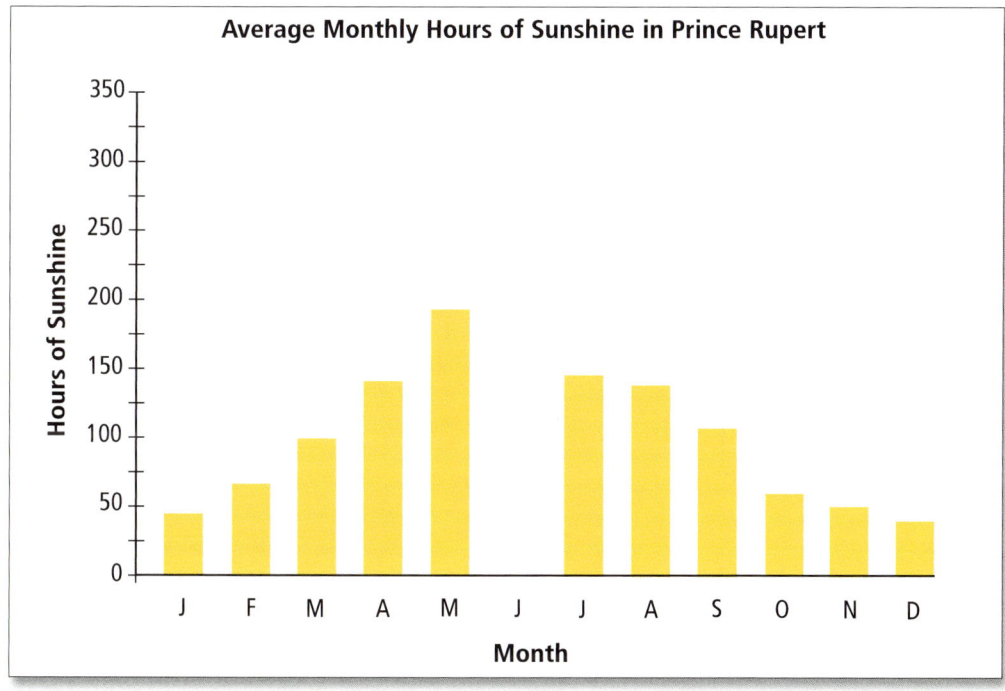

4. Predict what the number of hours of sunshine will be in Kamloops in May and in Prince Rupert in June. Using a copy of the bar graphs given to you by your teacher, shade these amounts in the graphs using a coloured pencil.

Lesson 10: Predicting From Patterns

What Did You Learn?

Meet with another pair of classmates to share, compare, and discuss your chart and graph predictions.

1. Which predictions seem most reasonable? Why?
2. In which month would you visit Kamloops if you wanted to have the greatest chance of enjoying sunny weather? Why?
3. In which month would you visit Prince Rupert if you wanted to have the greatest chance of enjoying sunny weather? Why?
4. How might you improve your predicting skills?

Practice

1. If you were making a weather prediction about how long winter might last, which of the data given below would you use to make the most accurate forecast?
   - whether or not Wiarton Willie saw his shadow
   - information from Environment Canada weather stations
   - recollections of farmers

   Discuss your answer with a classmate.

2. Review other tables and graphs of weather data from this chapter. Choose one table or graph in which the data shows a clear pattern. Write a prediction based on the data, and give reasons for your prediction. Then, meet with a classmate to share and compare predictions. Discuss how reasonable each prediction is and why it is reasonable.

Technology

The Old Farmer's Almanac is well known for making predictions about the weather. Access the Web site www.almanac.com/index.php and click on the weather link. Read the prediction for the weather in your area. Would you say the almanac's prediction is accurate? Explain why or why not.

Chapter Review

Chapter 2

MEASUREMENT
DATA MANAGEMENT AND PROBABILITY

1. Make the following metric conversions:
   a) 6 m = ■ cm
   b) 12 682 m = ■ km
   c) 2381 cm = ■ m
   d) 182 mm = ■ cm
   e) 1.2 km = ■ cm
   f) 80 dam = ■ m
   g) 5 dm = ■ m

2. Show these times on an analog clock face:
   a) 8:15 A.M.
   b) 12:31 P.M.
   c) 4:44 A.M.
   d) 7:31 A.M.
   e) 10:15 P.M.
   f) 4:55 P.M.
   g) 5:04 P.M.
   h) 7:36 A.M.

3. Write the following dates in SI notation:
   a) January 15, 1952
   b) July 1, 1867
   c) October 31, 2003
   d) a classmate's birthday

4. You've just received a gift of $50.00.
   a) Estimate combinations of items you can buy with your $50.00. Record your estimates.
   b) Check your estimates by adding. You may use a calculator to check your calculations.
   c) If you were to buy one of each of the items on the right with a $100.00 bill, how much change would you get back? Show your work.

$19.95

$24.95

Admit one
$14.00

$16.95

$15.95

Chapter Review 111

5. A class kept records of different types of weather in their area over a number of months. Their partial results are shown in the tally/frequency chart below.

Weather	Tally	Frequency
rain		5
snow	|||	
fog		2
high winds	++++	
hail			

a) Copy the above chart in your notebook. Provide all missing tally marks and frequencies.

b) What is the most frequent type of weather during this period?

c) What is the least frequent type of weather during this period?

6. A class recorded the following temperature highs for a school week during June:

Monday   22°C   Tuesday   21°C
Wednesday   23°C   Thursday   24°C
Friday   22°C

a) Construct a line graph to show the changes in temperature.

b) Write at least two conclusions you can draw from your line graph.

7. During a heavy snowfall, the following snowfall amounts were recorded at communities in Southern Ontario:

Hamilton   32 cm   Oakville   30 cm
Milton   31 cm   Mississauga 28 cm
Toronto   24 cm

a) Construct a bar graph to show the different snowfall amounts.

b) Write at least two conclusions you can draw from your bar graph.

Chapter 2: Weather Data

8. Southwestern Ontario has many lightning storms.

   The following number of lightning storms was noted in Southwestern Ontario communities in one year:

   Community A   9   Community B   16
   Community C   10   Community D   11
   Community E   13

   a) Construct a pictograph to show the number of lightning storms in these Ontario communities.

   b) Write at least two conclusions you can draw from your pictograph.

9. Find the mean of each of the following sets of data:
   a) 15, 19, 17, 23, 16
   b) 31, 35, 34, 30, 35
   c) 115, 130, 117, 110, 113

10. Predict each missing number in the following patterns. Write a sentence giving a reason or reasons to support each prediction.
    a) 8, 10, 12, 14, 12, 10, ■
    b) 2, 4, 8, 14, 22, 32, ■
    c) 20, 15, 11, 8, 6, ■

11. Ms. Petroni kept records of the amount she spent to have snow removed from the parking lot of her business each January. These are the amounts she recorded:

    1999 $630.00
    2000 $575.00
    2001 $380.00
    2002 $750.00
    2003 $645.00

    a) Which type of graph would be most appropriate for displaying this data? Why?

    b) Construct the graph you selected.

    c) Write at least two conclusions you can draw from your completed graph.

Chapter Review

Chapter Wrap-Up

Graph Weather Data

Monthly Snowfall in Churchill, Manitoba	
Month	Snowfall in Centimetres
January	16.9
February	14.6
March	18.6
April	22.3
May	19.5
June	3.5
July	0
August	0
September	6.4
October	29.3
November	41.6
December	22.8

You have reached the end of the second chapter in Unit 1. Throughout the chapter, you learned how mathematical ideas are related to describing and predicting weather, and how weather and climate affect people. You learned the following topics and skills: metric prefixes; reading and writing time according to the 12-h clock, and dates according to SI notation; reading and writing money amounts up to $1000.00; using a tally/frequency chart to organize data; reading and constructing line graphs, bar graphs, and pictographs; and making predictions based on data patterns.

Your task now is to work as part of a small team to create an appropriate graph to display data from a table. Then, you will write journal entries telling how snowy weather affects your life.

Chapter 2: Weather Data

1. Meet in a small group to discuss which would be the best graph to display the data in the table shown on the previous page. Make sure you can provide good reasons for your choice.

2. Use the data in the table to make the type of graph you have selected.

3. Meet with other groups to compare, share, and discuss your graphs. Use the following questions to start your discussion:

   a) Which graph constructed by the groups was best for displaying the snowfall data? Why?

   b) What patterns do you see in the graph data? How do the patterns compare with snowfall in your community? How might you show the comparison?

   c) What conclusions can you draw from this data?

4. Give feedback to others groups about their graphs. What are the strengths of each graph? What could be improved?

5. Calculate the mean for your snowfall data. What can you learn about your data from the mean?

6. Write a journal page covering a snowy winter day. Give the date in SI notation. Record the time of each entry according to the 12-h clock. In your entries, write about how weather affects you in the following areas:

   - clothing
   - transportation
   - recreation
   - housing
   - heating
   - costs

   Use numbers to provide details in your descriptions; for example, you might say, "I had to wear two layers of clothing" or "Our monthly electricity bill was $175.00!"

   Share your journal page with the class.

Chapter 3

Weather Patterns

In this chapter, you will learn the skills necessary to identify and explain weather patterns. At the end of the chapter you will create a group report based on important weather patterns in your local area.

In this chapter, you will
- identify, create, and extend patterns
- identify and explain pattern rules
- recognize the relationship between the position of a number and its value
- explain relationships between patterns
- solve and pose pattern problems
- make predictions based on patterns presented in tables
- find missing factors in equations
- explore how changing one factor affects a product or a quotient

With a small group, discuss patterns you see in your local weather.

Have fun exploring climate and weather patterns!

Lesson 1

Extending and Creating Patterns

WEATHER REPORT

PLAN:
You will learn to recognize patterns, explain the rules for patterns, and create patterns.

DESCRIPTION:
Climatologists study long-term weather conditions. They look for patterns that help us better understand and predict weather. For example, climatologists found that at night and in the morning, in the middle of Montréal, the temperature is sometimes five to six degrees warmer than in surrounding areas.

Get Started

A pattern repeats in a regular way. Patterns might appear in designs, colours, or numbers.

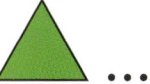

This pattern is a red square and a green triangle, which repeats. Each red square and green triangle is a sequence. If you continued the pattern for another sequence, you would add a red square and a green triangle.

Journal

Record examples of patterns you notice around you. Draw at least two sequences of the patterns. Then, extend the pattern one more sequence. Write a sentence giving the rule for each pattern.

Lesson 1: Extending and Creating Patterns

1. Work with a partner. For each row of items on the right, decide whether or not it is a pattern. Explain why.

A

B

C

2. Write a sentence that describes what the rule is for each pattern. What repeats? What sequence makes up the pattern?

3. Extend each pattern one sequence.

D **3, 6, 9, 12, …**

Meet as a large group or a class to share, compare, and discuss your observations and conclusions about patterns.

Vocabulary

pattern: Something, such as designs or numbers, repeated in regular ways

sequence: A succession of things, including numbers, that are connected in some way; for example, the sequence of numbers 2, 4, 6, 8, …

Build Your Understanding

Identify, Extend, and Create Patterns

Climatologists recorded the following temperature readings:
4°C, 8°C, 12°C, 16°C, …

Work in a small group to answer the first five questions.

1. What pattern do you notice in the temperatures?

2. If you continued the pattern, what would the next number be?

3. Explain how you decided this would be the next number.

4. Write a rule for the pattern.

Chapter 3: Weather Patterns

5. Write the next number for each of the following patterns:

a) 1, 3, 5, 7, ■

b) 1, 1, 2, 2, 3, 3, 4, ■

c) 1, 2, 4, 5, 7, 8, 10, ■

d) 1, 2, 4, 7, 11, 16, ■

6. Work on your own to do this exercise.

a) Create three number patterns. Make the first pattern easy, the second of average difficulty, and the third difficult.

b) Write at least three sequences in each pattern.

c) Write a sentence giving the rule for each pattern.

d) Meet with a classmate to take turns identifying each other's pattern. Give the next number in each pattern.

What Did You Learn?

1, 2, 2, 3, 3, 3, 4, 4, ■ , ■

1. With a classmate, give the next two numbers in the pattern.

2. How did you find the pattern rule?

3. In a sentence, describe the pattern rule.

Journal
Make notes on successful strategies you used for finding number patterns.

Practice

1. Use objects within the classroom to create patterns based on shape, colour, number, or some other characteristic. Write the rule for each pattern you create. Then, challenge a classmate to describe the next sequence in your pattern.

Lesson 1: Extending and Creating Patterns

Extension

2. In a small group, play a pattern game called I'm Going on a Vacation.

   a) One person is the tour organizer. He or she thinks of a pattern for things that those going on the vacation must bring. For example, the pattern might be all items that are orange, so the tour organizer begins the game by saying, "I'm going on a vacation and I'm bringing a basketball."

   b) The tour organizer then asks the first player what he or she will bring. If the item follows the pattern rule—a pumpkin, for example—the player can come. If not, he or she is not accepted for the tour. The tour organizer moves on to the next player.

   c) When a player thinks he or she knows the pattern rule, the player raises his or her hand and gives the rule. If it is correct, the player becomes the tour organizer and starts a new pattern. If not, the game continues until someone answers correctly.

Tip

- Be creative to make your patterns challenging.
- Listen carefully to what each player offers to bring. Look for patterns in who is allowed to go on the vacation and who is not.
- You might consider weather in your patterns. For example, bring things that protect you from the rain, such as umbrellas, hats, ponchos, and so on.

Technology

Use a paint application to create a collage of objects from our everyday lives that contain patterns. Use clipart, picture stamps, or images imported from the Internet to create the collage.

Booklink

How Math Works: 100 Ways Parents and Kids Can Share the Wonders of Mathematics by Carol Vorderman (The Reader's Digest Association, Inc.: Pleasantville, New York, 1996). This book features fascinating explanations, activities, and profiles of noted mathematicical thinkers. It also includes experiments that introduce readers to the world of math.

PATTERNING AND ALGEBRA
NUMBER SENSE AND NUMERATION

Lesson 2

Number Position and Value

WEATHER REPORT

PLAN:
You will use what you know about the position of a number and its value to solve and create number riddles about important weather facts.

DESCRIPTION:
People who study climate divide Canada into five climate regions: East Coast, Great Lakes, Prairie, West Coast, and Arctic/Sub-Arctic. Canada's hottest ever temperature was recorded in Saskatchewan.

A grain elevator on the Prairies

Get Started

Did you ever wonder what the hottest temperature in Canada was? Tomas knew this temperature record and challenged his friend Kenny to solve a riddle about the number. He gave these riddle clues:

The number has two digits.

It is between 30 and 50.

The tens digit can be evenly divided by 2, but the ones digit cannot.

The tens digit is 1 less than the ones digit.

1. Work with a partner to find the answer.

2. Meet with another pair to share, compare, and discuss your answers. As well, explain and discuss the strategies you used to find the mystery number.

3. What are some other clues Tomas might have used in his number riddle?

Vocabulary

place value: The value given to the place in which a digit appears in a numeral. For example, in 183, 1 is in the hundreds place, 8 is in the tens place, and 3 is in the ones place.

terms: The parts of an algebra expression separated by addition or subtraction signs. For example, $4x + 5$ has two terms, $4x$ and 5.

Lesson 2: Number Position and Value

Build Your Understanding

Solve and Create Weather Number Riddles

Work with a partner.

1. Medicine Hat, Alberta, has some of the warmest summers in Canada. Summer temperatures are recorded for June, July, and August. They are measured in degrees Celsius (°C). Use the following clues to find the average summer temperature of Medicine Hat.

    The number is between 20 and 30.

    Both the tens digit and the ones digit can be evenly divided by 2.

    You can multiply the tens digit by another number to obtain the ones digit.

    The difference between the tens digit and the ones digit is 4.

Tip
- Based on your knowledge of weather, estimate what you think the number will be. When you arrive at a riddle solution, decide whether or not your answer is reasonable.
- Use rough paper to experiment and narrow down number possibilities.
- Use a calculator to explore patterns and to check your work.

2. Farmers are especially interested in the number of hours of sunlight during June, July, and August. Medicine Hat has one of the sunniest summers in Canada. Use the clues to find the average number of hours of sunlight Medicine Hat receives in the summer.

    The number has three digits.

    It is between 900 and 1000.

    The hundreds digit can be evenly divided by 3.

    The tens digit is the smallest number, and the hundreds digit is the largest.

    The tens digit is 2 less than the ones digit.

    The ones digit is 2 less than the hundreds digit.

3. With your partner, choose three weather facts from the list below that interest you. Create a riddle with good clues for each fact you selected. Challenge another pair to solve your weather riddles.

- The hottest place in North America is Death Valley, California. Daily summer temperatures may stay above 49°C for over a month.
- The highest temperature ever recorded was 58°C in Libya in 1922.
- The place in the world with the hottest annual temperature in the shade is Dalol, Ethiopia. The average temperature there is 34.4°C.
- The place in Canada with the most hours of sunlight for June, July, and August is Yellowknife, with 1037 h.
- In the Sahara Desert, temperatures frequently soar to more than 50°C.

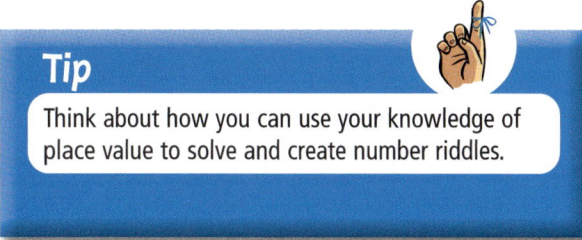

Tip

Think about how you can use your knowledge of place value to solve and create number riddles.

What Did You Learn?

1. Which riddle clues did you find least helpful? Why? How could you improve the clues?
2. Which riddle clues did you find most helpful? Why?
3. Goose Bay holds the Newfoundland and Labrador record for the most hours of sunshine, at 1806 h.
   a) What is the position of the 8?
   b) What is its value?
   c) How are the position and value of the 8 different from the position and value of the 1?

Lesson 2: Number Position and Value

Practice

Extension

1. The warmest temperature in Québec was recorded in Ville Marie on July 6, 1921. Solve this riddle to find the record.

   The number has two digits and is between 35 and 45.

   The tens digit can be evenly divided by 2.

   The entire number can be evenly divided by 5 and by 10.

2. Create a math/weather game.

   You Will Need
   - a Snakes and Ladders board game
   - markers and number cubes
   - file cards to make game math/weather cards

   a) Work with the same partner and pair as you did for question 3 in Build Your Understanding. Pool all the weather fact riddles created by the members of the group from question 3, and write each one on a game card. You will need to research additional weather facts and create more weather riddles to make a large enough pack of game cards.

   b) Use your new game cards when you play Snakes and Ladders. For example, when a player lands on the end of a snake or a ladder, he or she must solve a math/weather riddle card. If the riddle is answered correctly, the player moves up the snake or ladder. If the riddle is answered incorrectly, the player moves down the snake or the ladder.

   c) The first player to reach "Finish" becomes a class climatologist.

PATTERNING AND ALGEBRA

Lesson 3
Number Pattern Relationships

WEATHER REPORT

PLAN:
You will practise identifying number patterns, discuss pattern rules with classmates, and also discuss relationships between patterns.

DESCRIPTION:
All places have a climate, including your classroom. Inside temperatures can have patterns that vary, depending on where you measure them within the classroom. What would be the warmest place within the classroom? Why? What would be the coldest place? Make sure to record your predictions.

Get Started

Sometimes two or more weather patterns can be related.

Dan lives on the outskirts of a Canadian city. Each summer he spends a number of weeks at his grandparents' cottage in the country. Before leaving, Dan watches five-day weather forecasts for the city and the lake where his grandparents' cottage is located. He wants to know how weather might affect activities he plans.

Lesson 3: Number Pattern Relationships

Table A: Temperatures Forecasted for Dan's Home Near the City				
Monday	Tuesday	Wednesday	Thursday	Friday
28°C	29°C	31°C	32°C	34°C

Table B: Temperatures Forecasted for His Grandparents' Cottage at the Lake				
Monday	Tuesday	Wednesday	Thursday	Friday
26°C	27°C	29°C	30°C	32°C

Work with a classmate or in a small group.

1. What is the rule for the pattern in Table A?
2. What is the rule for the pattern in Table B?
3. How are the two patterns related?
4. Predict the temperature for Saturday if the pattern in Table A continues.
5. Predict the temperature for Saturday if the pattern in Table B continues.
6. Meet in a group or as a class to compare, share, and discuss your conclusions, predictions, and prediction strategies.

Journal

Make notes on any related number patterns you observe. They could be about weather or another area that interests you, such as sports statistics. Explain the rule for each pattern, and write a sentence explaining how the patterns are related.

Vocabulary

table: A useful tool for organizing and presenting information

Chapter 3: Weather Patterns

Build Your Understanding

Find Patterns and Relationships Between Patterns

In their science work, Mr. Grabowski's class studied energy efficiency. They decided to find the energy efficiency of their own classroom. They measured classroom temperature near the floor, halfway between the floor and ceiling, and near the ceiling. Each measurement was taken at mid-afternoon. The class organized their results in this table.

Temperatures in Mr. Grabowski's Class			
Day of the Week	A: Floor	B: Halfway	C: Ceiling
Monday	22°C	23°C	24°C
Tuesday	22°C	23°C	24°C
Wednesday	23°C	24°C	25°C
Thursday	23°C	24°C	25°C
Friday	24°C	25°C	26°C
Saturday			

1. Work in small groups and help one another to read the table. What patterns do you notice in table information?

Tip
Use features, such as the title and headings, to help understand the table. Look for patterns in table columns, then for patterns in rows.

2. What is the pattern rule for Column A: Floor?
3. What is the pattern rule for Column B: Halfway?
4. What is the pattern rule for Column C: Ceiling?
5. How are the patterns in Columns A, B, and C related?
6. Predict what the Saturday temperature would be based on this pattern.
7. Share, compare, and discuss your conclusions with another group or the entire class.

What Did You Learn?

1. At the beginning of the lesson, you predicted where the temperatures would be warmest and coolest. How do your predictions compare with the results of Mr. Grabowski's class?

2. The class also recorded the temperature on a thermometer outside the classroom (outdoors).

Table of Temperatures Outside Mr. Grabowski's Class					
Monday	Tuesday	Wednesday	Thursday	Friday	Saturday
14°C	14°C	15°C	15°C	16°C	

   a) What is the pattern rule in their results?
   b) Predict what the temperature would be for Saturday based on this pattern.
   c) How is this pattern related to results the class obtained for temperatures inside the classroom?

3. If a student turned the classroom thermostat up 3°C, what effect would this change have on the temperatures of Columns A, B, and C in the chart on the previous page? Explain your prediction.

4. Explain how tables can help you do the following:
   a) organize information
   b) see patterns
   c) see relationships between and among patterns

Practice

Math Problems to Solve

1. Create a pattern made up of six whole numbers. Create a second pattern of six whole numbers that is related to your first pattern. Write a sentence to explain how the two patterns are related.

2. Exchange your patterns with a classmate, and challenge him or her to explain how the patterns are related.

Show What You Know

Review: Lessons 1 to 3, Patterns

1. Write the next number for each of the following patterns. Describe the pattern rule for each.
   a) 5, 10, 15, 20, ■
   b) 10, 20, 40, 70, 110, ■
   c) 1, 3, 7, 13, 21, 31, ■
   d) 1, 5, 9, 13, 17, 21, ■

2. Use the clues below to find the average daily high temperature for Montréal, Québec, in July.

   The number is between 10 and 30.

   The tens digit and the ones digit can be evenly divided by 2.

   The sum of the two digits is more than 6 but less than 10.

3. A class recorded the temperatures in the charts below outside and inside their classroom. Work with a classmate to do the following:
   a) Identify the pattern of temperatures outside the classroom.
   b) Identify the pattern of temperatures inside the classroom.
   c) Explain how the patterns are related.
   d) Predict what the temperatures outside and inside the classroom will be on Saturday if the patterns continue.

Table of Temperatures Outside the Classroom					
Monday	Tuesday	Wednesday	Thursday	Friday	Saturday
22°C	23°C	25°C	26°C	28°C	

Table of Temperatures Inside the Classroom					
Monday	Tuesday	Wednesday	Thursday	Friday	Saturday
18°C	19°C	21°C	22°C	24°C	

NUMBER SENSE AND NUMERATION
PATTERNING AND ALGEBRA

Lesson 4

Multiplication and Division Patterns

WEATHER REPORT

PLAN:
You will explore patterns in dividing by 10 and 100 and multiplying by 0.1 and 0.01. You will use your knowledge of these pattern rules to solve problems.

DESCRIPTION:
You might hear on a television weather broadcast or read in a newspaper about weather comparisons. For example, one Canadian community receives annual rainfall of 1455 mm, while a second community receives $\frac{1}{10}$ that amount. Think about how you would find the annual rainfall in the second community.

Get Started

Tavia and Anil learned that a mountain area near Revelstoke, British Columbia, once received 2447 cm of snow in a year. Another location in Canada received $\frac{1}{10}$ that amount of snow. They each had a different method for finding the amount of snow the second location received.

Tavia's Division Method

Tavia used division. She divided 2447 by 10 to get the answer.

Tavia knows that you can divide numbers by 10 mentally. All you have to do is move the decimal point in the number one place value or space to the left. The answer (quotient) is 10 times smaller. If the number being divided by 10 is a whole number, as in this question, the decimal point comes after the ones digit.

$$2447. \div 10 = 244.7$$

The second location in Canada received 244.7 cm of snow.

1. Why do you think Tavia divided by 10? Discuss your answer with the class.

Anil's Multiplication Method

Anil multiplied to find the answer. He multiplied 2447 by 0.1.

Anil knows that you can multiply numbers by 0.1 mentally. All you have to do is move the decimal point in the number one place value or space to the left. The answer (product) is 10 times smaller. If the number being multiplied by 0.1 is a whole number, as in this question, the decimal point comes after the number in the ones digit.

$$2447. \times 0.1 = 244.7$$

The second location in Canada received 244.7 cm of snow.

2. Why do you think Anil multiplied by 0.1? Discuss your answer with the class.

Work with a classmate to complete the following:

3. Use a calculator to check Tavia's and Anil's work.

4. How are the two methods related?

5. Community A annually gets 8123 mm of rainfall. Community B gets $\frac{1}{100}$ of that amount. Use the division and multiplication methods to find out how much rainfall Community B receives each year.

6. Use a calculator to check your work.

7. Write a rule for dividing a number by 100 and multiplying a number by 0.01. Use the information on the previous page to help you.

8. Meet with another pair of classmates to share and discuss your answers and calculation methods.

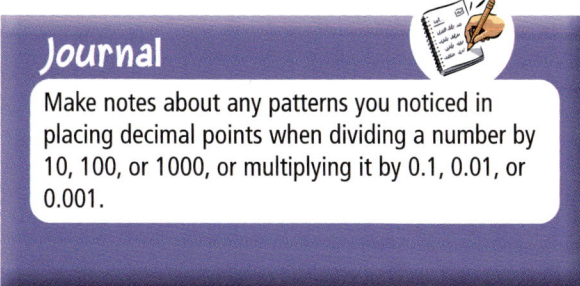

Journal

Make notes about any patterns you noticed in placing decimal points when dividing a number by 10, 100, or 1000, or multiplying it by 0.1, 0.01, or 0.001.

Build Your Understanding

Divide by 10 and 100 and Multiply by 0.1 and 0.01

1. **a)** Divide the following numbers by 10. Write all answers as decimal numbers.

    80   875   67.8   5.32

    Tip: Use a calculator to check your work.

   b) Multiply each number in 1(a) by 0.1.

   c) Describe any relationships you noticed between dividing by 10 and multiplying by 0.1.

2. **a)** Divide the following numbers by 100. Write your answers as decimal numbers.

    700   628   54   93.6

   b) Multiply each number in 2(a) by 0.01.

   c) Make note of any relationships you noticed between dividing by 100 and multiplying by 0.01.

3. Victoria, British Columbia, receives about $\frac{1}{10}$ of the average annual snowfall of Sept-Îles, Québec. Sept-Îles receives 415 cm. How much does Victoria receive? Use two methods to find the solution.

   Journal: Which method did you find most effective for answering questions and solving problems in this activity—dividing by 10 and 100, or multiplying by 0.1 and 0.01? Write a sentence or two explaining why.

What Did You Learn?

1. How can you divide by 10 and 100, or multiply by 0.1 and 0.01, without using a calculator?

2. Meet with a small group of classmates to share, compare, and discuss your answers for questions 1 to 3 in Build Your Understanding, the calculation strategies you used, and any patterns you observed.

Practice

Extension

Divide these numbers by 10 and 100.

1. 248   **2.** 586   **3.** 999   **4.** 1234

5. What pattern do you notice?

6. Write a rule for dividing a decimal number by each of 10 and 100.

7. Check your rule by testing other numbers.

Multiply these numbers by 0.1 and 0.01.

8. 17.2   **9.** 284   **10.** 662   **11.** 1548

12. What pattern do you notice?

13. Write a rule for multiplying a decimal number by 0.1 and 0.01.

14. Check your rule by testing other numbers.

Lesson 4: Multiplication and Division Patterns

Lesson 5

Patterns With Variables

WEATHER REPORT

PLAN:
You will learn what variables are, and identify and create patterns with one and two variables. Then you will use pattern rules with variables, called "formulas," to solve problems.

DESCRIPTION:
At Canadian weather stations, weather scientists collect information about patterns. These patterns help the scientists make more accurate predictions, or forecasts. If you look closely, you can also find patterns in the geometric shapes of some weather stations.

PATTERNING AND ALGEBRA
MEASUREMENT
NUMBER SENSE AND NUMERATION

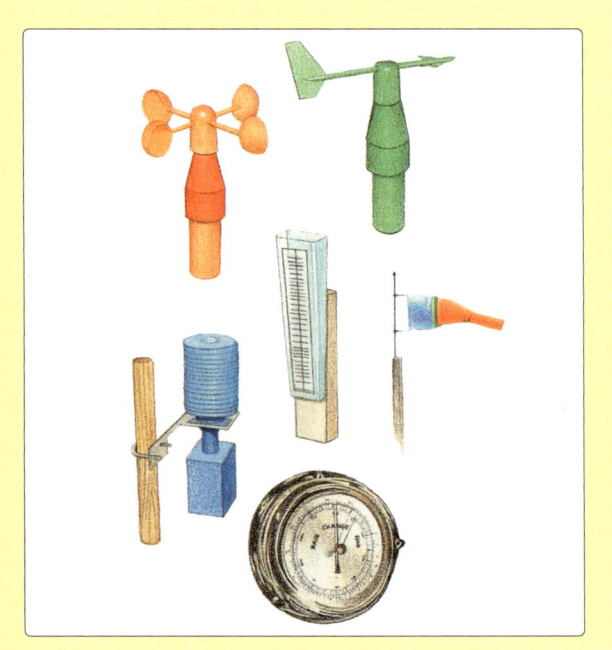

Get Started

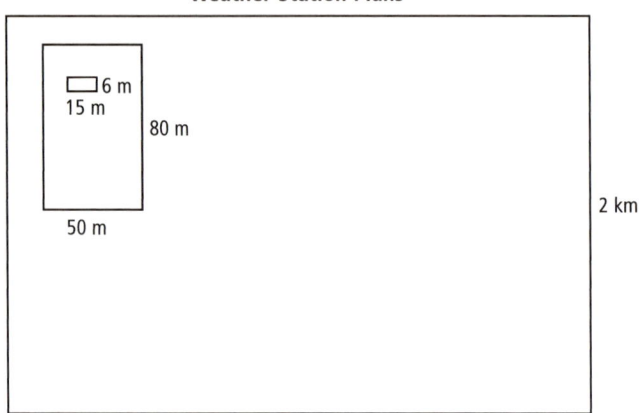

Weather Station Plans

The rule, or formula, for finding the perimeter (distance around) a rectangle is 2l (length) + 2w (width).

Here is the formula broken down:

Perimeter = 2 lengths + 2 widths

or $P = 2 \times l + 2 \times w$. This is usually written as $P = 2l + 2w$.

Chapter 3: Weather Patterns

The formula for the perimeter of a rectangle includes the variables *l* and *w*. Variables use letters that represent numbers that change. For example, in the rectangles on the previous page, *l* can be 6 m, 80 m, or 2 km.

The number 2 in the perimeter formula is a constant. It stays the same for all rectangles. In a rectangle, there are always 2 side lengths and 2 side widths.

1. Work with a partner. What numbers can the width variable stand for in the weather station diagram? Why?

2. You can also find variables in patterns.
   2, 5, 8, 11, 14, …
   a) What is the next number in the pattern?
   b) What is the amount by which each number increases in the pattern? This is the constant difference.
   c) The constant is used to create the other terms (numbers) in the pattern. These terms keep changing. They vary. Choose a variable to represent these changing numbers.
   d) What is the pattern rule? Write it using the variable you selected.

Vocabulary

algebra: A branch of mathematics that uses expressions, variables, constants, and other symbols to make mathematical statements

constant: A term in an algebraic expression that does not change. For example, in $x + 3$ the constant is 3.

dimension: A measurement such as the height, width, length, or depth of an object

formula: A set of symbols, such as letters and numbers, used to give a general rule. For example, the formula for finding the perimeter of a rectangle is $P = 2l + 2w$; the formula for finding the area of a rectangle is $A = l \times w$.

terms: The parts of an algebraic expression, separated by addition or subtraction signs. For example, $4x + 5$ has two terms, $4x$ and 5.

variable: A letter or symbol used to represent a number. A variable changes; for example, in the formula $2l + 2w$, *l* and *w* are variables.

Tip
Use variables to explain the pattern rule.

Lesson 5: Patterns With Variables

Build Your Understanding

Identify Variables in a Pattern

You Will Need
- grid paper

1. Work with a classmate to study this set of numbers.

   7, 14, 21, 28, 35, ….

   a) What is the next number in the pattern?

   b) What is the amount by which each number increases in the pattern?

   c) What is the constant in the pattern? Select a variable and write the pattern rule.

2. Create a different pattern that changes using addition or subtraction.

3. Challenge another pair of classmates to give the next two numbers in your pattern and to explain the pattern rule using a variable and a constant.

4. Use the two variables in the perimeter formula to find perimeters of each rectangle in the weather station plans from Get Started.

5. On grid paper, draw plans for a weather station. Include at least three rectangles in your plans. Give the length and width of all shapes in the plans. Use the correct formula and variables to find the perimeters of all figures in your drawing. You might label where in your station plans you would place important weather instruments, such as weather vanes and precipitation gauges.

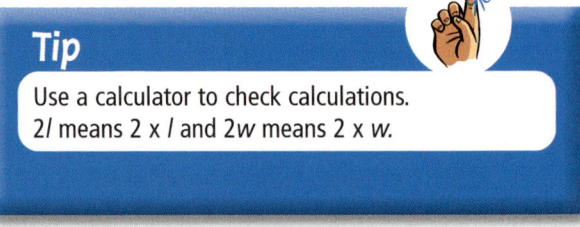

Tip

Use a calculator to check calculations. $2l$ means $2 \times l$ and $2w$ means $2 \times w$.

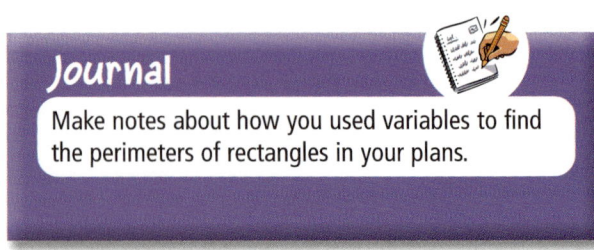

Journal

Make notes about how you used variables to find the perimeters of rectangles in your plans.

What Did You Learn?

The formula for finding the area of a rectangle is
Area = length x width or $A = l \times w$
This is usually written as $A = lw$

1. What are the variables in the formula?

2. Are there any constants? If so, what are they?

3. Use the area formula to find the areas of rectangles in this lesson.

4. Explain what formula you would use to find the perimeter of a square. What are the variables in the formula? Are there any constants? If so, what are they?

5. Explain what formula you would use to find the area of a square. What are the variables in the formula? Are there any constants? If so, what are they?

Practice

Extension

You learned that a variable is a letter or symbol used to represent a number. Often, people use letters in place of numbers to remember important telephone numbers. For example, an information number about tornadoes might be

(555) BIG–WIND for (555) 244–9463

1. Create at least three weather-related phone numbers in which you replace numbers with letters.

2. Challenge classmates to find which number each letter in your proposed telephone number represents.

Lesson 5: Patterns With Variables

Lesson 6

Exploring Pattern Rules

WEATHER REPORT

PLAN:
You will learn pattern rules that relate to weather, and use these rules to solve number problems. You will also explore and discuss relationships between patterns.

DESCRIPTION:
By studying weather over long periods of time, scientists can see patterns that help them better understand weather. For example, they know that climate and weather are affected by landforms, such as mountains; water, such as oceans and large lakes; wind; how far a place is north or south of the equator; and ocean currents.

Get Started

Algebra is a strand of mathematics that uses expressions, variables, constants, and other symbols to make mathematical statements. Algebra is helpful when working with weather patterns. For example, during the winter in Winnipeg, Manitoba, the temperature at the centre of the city is usually 3°C to 4°C warmer than the temperature at the airport.

Chapter 3: Weather Patterns

Winnipeg Airport

Downtown Winnipeg at Portage and Main

Work with a partner or in a small group.

1. What pattern do you notice in each column of the table?
2. How are the patterns related?
3. What is the pattern rule for Winnipeg winter temperatures at the airport and downtown?

Winnipeg Temperatures	
Temperature at the Airport	Temperature Downtown
1°C	5°C
2°C	6°C
3°C	7°C
4°C	8°C
5°C	9°C

An algebra statement, or rule, about the relationship between the two temperature patterns would be:
Temp. downtown (T)

= Temp. airport (x) + 4
= $x + 4$

The variables are T and x and the constant is 4.

Vocabulary

algebra: A branch of mathematics that uses expressions, variables, constants, and other symbols to make mathematical statements

4. If the temperature at the airport is 0°C, what will the downtown temperature be?
5. If the temperature at the airport is 8°C, what will the downtown temperature be?
6. If the temperature downtown is 10°C, what will the temperature at the airport be? Explain to a classmate how you found the answer.

Lesson 6: Exploring Pattern Rules

Build Your Understanding

Use Tables and Apply Rules to Solve Problems

Work with a classmate or in a small group to solve each problem.

> **Tip**
> Draw a labelled diagram to help you better understand the problem.

1. Weather scientists discovered this rule: at higher parts of the mountains, the air cools about 1°C for every 200 m it rises. If the air is 15°C and rises 800 m, what will the new temperature be?

Changing Mountain Temperatures			
Temperature at Start	Amount Air Rises	Change in Temperature	New Temperature
15°C	200 m	1°C	14°C
15°C	400 m	2°C	13°C

2. With your partner or in a small group, study the chart above. What patterns do you see developing?

3. How could you use these patterns and the chart to help you solve the problem?

4. Copy the chart into your notebook. Complete all chart entries needed to solve the problem.

5. As a group, write an algebra statement with a variable and a constant that lets you calculate a new temperature when mountain air rises. Use the statement to check your solution to the problem.

6. Meet with another group, or as a whole class, to share, compare, and discuss the algebra statement you used to find new temperatures. Identify all variables and constants in the statement, and explain why you used them.

> **Journal**
> If necessary, improve your algebra statement based on new ideas you learned from sharing and discussing statements with classmates. Try out the "new and improved" algebra statement by using it to solve the problem from Build Your Understanding.

What Did You Learn?

1. Explain how developing a table helped you solve the problem.

2. Explain to a classmate how developing and using an algebra statement helped you solve the problem.

3. Which method did you find more useful? Why?

Practice

Math Problems to Solve

Weather scientists have found that, in snowbelt areas, a rise of 30 m in the height of the land adds about 17 cm in the yearly average snowfall. If a community near the lake has a yearly snowfall of 115 cm, what will the yearly snowfall be for an area that has a height of 90 m?

Some areas in Southern Ontario receive more snow than do others. Some reasons for this are closeness to major lakes and the height of the land.

1. Develop a table to solve the problem.

2. Then, develop an algebra statement to solve the problem. Use the statement and calculations to check the solution you obtained with the chart.

3. Meet with a classmate to share, compare, and discuss your solution and problem-solving strategies.

Lesson 6: Exploring Pattern Rules

Show What You Know

Review: Lessons 4 to 6, Patterns

1. Divide each number by 10 and 100:
   a) 3000   b) 764   c) 254.1

2. Multiply each number by 0.1 and 0.01:
   a) 6000   b) 890   c) 370.2

3. Create patterns that follow these pattern rules:
   a) goes up by one, then by two
   b) goes up by two, then by one
   c) goes up by three, then by two

4. Research weather rules. Use the rules to create interesting mathematical problems that require classmates to develop tables or use algebra statements to find the solutions.

5. Find the missing variables to make these equations true:
   a) $175 = 5x$
   b) $12x = 144$
   c) $20x = 400$

Lesson 7
Finding Missing Factors

WEATHER REPORT

PLAN:
You will learn how to find a missing factor in an equation. You will also decide whether a mathematical sentence is always, sometimes, or never true.

DESCRIPTION:
Clusters of twisters, also known as tornadoes, can travel very quickly, leaving a terrible path of destruction.

Get Started

An equation is a mathematical statement that has equivalent, or equal, terms on the two sides of the equal sign (=); for example:

3 x 5 = 15

When solving multiplication questions, you multiply two or more numbers called factors to obtain a product.

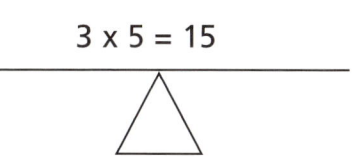

3 x 5 = 15
factors   product

Lesson 7: Finding Missing Factors

You might sometimes see an incomplete equation with a missing factor, so that the equation is not balanced, such as:

$4 \times ? = 12$

To find the missing factor, or what "?" is, divide 12 by 4. The missing factor is 3.

$4 \times 3 = 12$

The equation is now complete and balanced. Terms on both sides of the equal sign are equivalent.

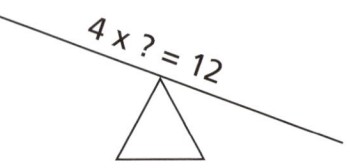

Journal

Write a sentence or two explaining how you find missing factors. Include an example with your explanation.

Vocabulary

equation: A mathematical statement that has equivalent terms on either side of the equal sign (=). For example, $2 \times 3 = 6$.

equivalent: Equal to; the same amount

factor: A number that divides evenly into another number or expression. For example, 4 is a factor of 12 because $3 \times 4 = 12$.

Build Your Understanding

Find Missing Factors

1. Find the missing factors in the following equations:

   a) $6 \times \square = 42$ b) $\square \times 8 = 96$ c) $12 \times \square = 144$

   d) $\square \times 12 = 108$ e) $11 \times \square = 132$ f) $9 \times \square = 72$

   g) $10 \times \square = 1000$ h) $100 \times \square = 10$ i) $10 \times \square = 0.01$

Tip

Review your basic multiplication facts.

2. What would be the first thing you would do to find the missing factor in this equation?

   $9a = 45$

   Explain to a classmate why you did this first. Follow your method to find the missing factor.

Chapter 3: Weather Patterns

3. Work with a classmate to find the missing factor "x" and balance these equations.

   a) $2x = 16$ b) $6x = 36$ c) $7x = 63$
   d) $125 = 5x$ e) $15x = 225$

Tip
- Use a calculator to help you find missing numbers.
- The rule $2y = 16$ has y as the variable. This means the rule is $2 \times y = 16$.

Work in a small group to solve the following problems.

4. The average twister travels at a speed of 45 km/h. In 1974, a twister traveling at a speed of 120 km/h was recorded in Alabama. If it travelled 480 km in total, how many hours did the twister travel?

   a) Write the problem as a multiplication equation with a missing factor.
   b) Use what you know about finding missing factors to solve the problem.

5. During the St. Lawrence River Valley ice storm in 1998, 100 mm of freezing rain fell over 6 days. On average, how much fell each day?

   a) Write the problem as a multiplication equation with a missing factor.
   b) Use what you know about finding missing factors to solve the problem.

Lesson 7: Finding Missing Factors

What Did You Learn?

1. How did mental mathematics help you find missing factors in this lesson? Write one question in which mental mathematics was especially useful.

2. Explain to a classmate how you can use missing-factor equations to solve problems. Give an example weather problem in your explanation.

3. How does the relationship between division and multiplication help you find the value of a missing factor in equations involving multiplication?

4. **a)** Write all the factors of 8 from least to greatest. Then, write all the factors of 12 from least to greatest.

   b) What is the least factor 8 and 12 share? What is the greatest factor they share?

Practice

1. Find the value of y that makes each equation true.
   a) $4y = 44$
   b) $3y = 18$
   c) $3y + 4 = 19$

2. Which of the following equations are true? Which are false?
   a) $7 \times 6 = 63$
   b) $4 \times 12 = 48$
   c) $9 \times 7 = 71$
   d) $5 \times 15 = 75$

PATTERNING AND ALGEBRA
NUMBER SENSE AND NUMERATION

Lesson 8

Changes in Factors

> **WEATHER REPORT**
>
> **PLAN:**
> You will explore how changing one factor in a multiplication or division problem affects a product or quotient.
>
> **DESCRIPTION:**
> Using a simple barometer is the most reliable way for most people to predict storms. A sudden drop in air pressure means a storm is coming.

A barometer

Get Started

Multiplying

Barometers measure changes in air pressure. You can also see change in a multiplication product when you change one factor.

$$7 \times 5 = 35$$
Factor Factor Product

1. Work with a classmate. Predict what will happen to the product if you increase the first factor by 1.

2. Test your prediction by multiplying $8 \times 5 = $ ▇. How accurate was your prediction?

3. What will happen to the product if you increase the first factor again by 1? Multiply to test your prediction.

4. What multiplication pattern do you see developing? State the pattern as a rule. Add 1 to the first factor and multiply a few more times to test whether your pattern rule holds.

5. What will happen if you increase the second factor (5) by 1? Multiply to test your prediction.

6. Continue adding 1 to the second factor and then multiplying until you can state a pattern rule.

Lesson 8: Changes in Factors

Dividing

35 ÷ 7 = 5
(Dividend) (Divisor) (Quotient)

7. How are the divisor and quotient in the division question related to the factors in the multiplication question?

8. Predict what will happen if you divide the following. Then answer the questions.

   40 ÷ 8 = ■

   45 ÷ 9 = ■

9. Make up three more related division questions to extend the pattern. Answer the questions.

10. Meet with a group or as a whole class to share, compare, and discuss your answers and patterns you observed.

Journal
Make notes about any other important patterns you noticed in multiplication and division facts.

Vocabulary
dividend: The number being divided in a division calculation. For example, in 36 ÷ 6, the dividend is 36.
divisor: The number you are dividing the dividend by. For example, in 15 ÷ 3, the divisor is 3.
product: The answer in a multiplication calculation
quotient: The answer in a division calculation. For example, in 24 ÷ 4 = 6, the quotient is 6.

Build Your Understanding

Explore How Changing a Factor Can Affect a Product or Quotient

1. Work with a partner to complete the table.

Factor	2	3	4	5	6	7	8	9
Factor	9	9	9	9	9	9	9	9
Product	18		36			63		

2. What pattern do you notice in the product when you increase one of the factors by 1?

Chapter 3: Weather Patterns

3. Work with your partner to complete this table.

Factor	2	3	4	5	6	7	8	9
Factor	12	12	12	12	12	12	12	12
Product	24							108

4. In the table below, predict what the product will increase by in each table cell. Then complete the table to check your predictions.

Factor	2	3	4	5	6	7	8	9
Factor	124	124	124					
Product								1116

5. Use division to find the missing factors in the table below.

Factor	2	3	4	5	6	7	8	9
Factor								
Product	96	144	192	240	288	336	384	432

6. What pattern do you notice in the product? Extend the pattern by one number.

7. What effect will doubling the first number in the first factor pattern have on the product? How could you test your prediction?

What Did You Learn?

1. Copy the factor/product table below into your notebook. Create a factor pattern in which one factor is increased by 1. Keep the second factor constant. Enter your factor pattern in the table you drew in your notebook, leaving missing factors. Provide a few products as clues for a classmate to find the pattern.

Factor								
Factor								
Product								

2. Exchange your incomplete factor/product table with a classmate. Challenge him or her to complete the table.

3. Meet with other classmates to discuss completed charts.

Practice

1. Work with a classmate to find the missing factor "x" to balance each of these equations:

   a) $3x = 27$　　　b) $6x = 42$　　　c) $8x = 56$

   d) $121 = 11x$　　e) $16x = 128$

2. Which of the following equations are true? Which are false? Why?

   a) $5 \times 15 = 70$　　b) $9 \times 20 = 160$

   c) $152 \div 4 = 38$　　d) $1000 \div 10 = 10$

3. Estimate, then calculate the mean of each of the following sets of numbers:

   a) 28, 25, 25

   b) 47, 49, 53, 71

   c) 157, 159, 163, 295, 156

   d) 1481, 1483, 1485, 1491

Show What You Know

Review: Lessons 7 to 8, Factors

In the table below, predict what the product will increase by in each table cell. Then, complete the table to check your predictions.

Factor	2	3	4	5	6	7	8	9
Factor	50	50	50					
Product								450

Chapter 3: Weather Patterns

Lesson 9

Three-Dimensional Patterns

PATTERNING AND ALGEBRA
GEOMETRY AND SPATIAL SENSE

WEATHER REPORT

PLAN:
You will create three-dimensional patterns using coloured blocks. You will also use tiling patterns to cover a plane.

FACTS:
Weather forecasters use three-dimensional computer-model images to show the many forces that affect and create weather. The computer model in Get Started shows ocean circulation in the North Atlantic. The dark parts represent land.

Get Started

With a classmate, describe the patterns you see in the different clouds in the pictures above. How are the clouds similar? How are they different?

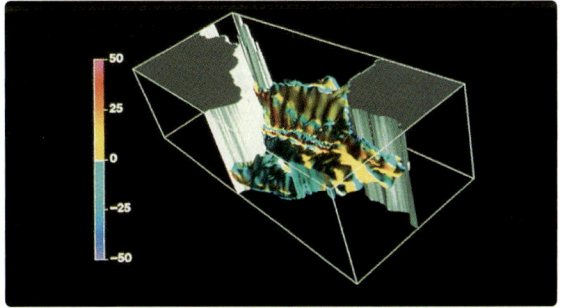

This three-dimensional computer model shows ocean circulation in the North Atlantic.

Lesson 9: Three-Dimensional Patterns   **151**

You may have seen this pattern on chess or checkers game boards. It is an example of a two-dimensional pattern. "Two-dimensional" means something has length and width, but not height.

1. Work in a small group. What is the rule for this two-dimensional pattern?

2. Find an example of a two-dimensional pattern in your classroom. Explain why it is two-dimensional.

3. What is the rule for the two-dimensional pattern you found?

4. Explain how you would extend the pattern by following the rule.

You Will Need
- grid paper
- black and red crayons
- black and red cubes

Patterns can also be found in three-dimensional shapes. A three-dimensional shape has length, width, and height.

5. What is the pattern rule for the three-dimensional pattern on the right?

6. Predict what the pattern would look like if you added another layer. On grid paper, draw your prediction.

7. With black and red cubes, copy the pattern shown in the picture. Add a new layer to the three-dimensional pattern. How accurate was your prediction?

Vocabulary

three-dimensional: Having three dimensions—length, width, and height

two-dimensional: Having two dimensions—length and width, but not height

Journal

In your own words, write definitions of "two-dimensional" and "three-dimensional." Include labelled diagrams to support your definitions.

152 Chapter 3: Weather Patterns

Build Your Understanding

Construct Two-Dimensional and Three-Dimensional Patterns

You Will Need
- grid paper
- black and red coloured pencils (or any two colours)
- black and red cubes (or any two colours)

1. Work in a small group. Use coloured cubes to make an interesting pattern that is one layer in height.

2. On grid paper, draw and colour a two-dimensional picture of the top view of your pattern.

3. Write sentences explaining the pattern rules for the two-dimensional and three-dimensional patterns you created.

4. Add a new layer of cubes following the pattern in the first layer.

5. Describe any new patterns you see developing.

6. Predict what patterns will develop if you extend the pattern one more layer. Show your prediction by making a coloured drawing of the three-dimensional object on grid paper.

7. Add the new layer of cubes to test your prediction. How accurate was your prediction?

8. Construct other more complex and decorative patterns with coloured cubes.

9. Challenge other groups to identify the pattern rules for any two-dimensional and three-dimensional patterns your group created.

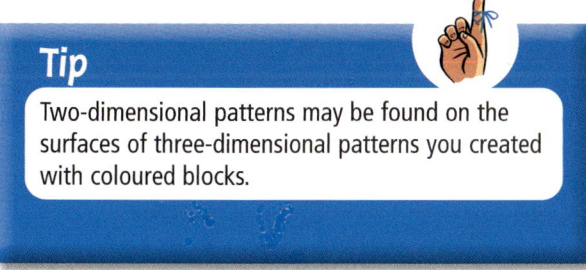

Tip

Two-dimensional patterns may be found on the surfaces of three-dimensional patterns you created with coloured blocks.

Lesson 9: Three-Dimensional Patterns

What Did You Learn?

1. Identify good examples of three-dimensional patterns in your classroom, around your school, or at home.

2. Explain why they can be called "three-dimensional."

3. Write a sentence clearly explaining the rule for each pattern.

4. Explain how you would use the rule to extend each pattern.

5. Meet with a group or your whole class to share, compare, and discuss the three-dimensional patterns you found.

Practice

Extension

You Will Need
- scissors
- grid paper
- coloured construction paper

1. As a group, draw a figure on grid paper. Make congruent copies of the figure.

2. Then, work independently to create and cut out different shapes to tile the plane. Next, use the shapes to tile the plane.

3. Meet as a group to share and compare your tiling patterns. What different tiling patterns did other group members create?

4. Are any of the tiling patterns also tessellations? How do you know?

5. Display your tiling patterns on a class bulletin board.

Vocabulary

congruent: Exactly the same size and shape
plane: A flat or level surface that extends without end in all directions
tessellation: A repeating pattern of closed figures that covers a surface with no gaps or overlaps
tile: To cover a plane surface with a combination of different shapes with no gaps or overlaps

Technology

Use a draw application to explore which geometric shapes are good tessellating tiles. Copy and paste the shapes to tile the work area to make an interesting pattern that tessellates.

Chapter Review

1. Extend each pattern, then explain the pattern rule.
   a) 2, 4, 6, 8, ▪   b) 1, 2, 2, 3, 3, 3, ▪   c) 2, 4, 16, 32, ▪
   d)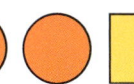

2. Solve the following number riddle:
   - The number has three digits.
   - Two of the digits are even numbers and one is odd.
   - The hundreds digit divides evenly into the ones digit.
   - The number is less than 260 and greater than 250.
   - A number squared (times itself) equals this number.

3. The weather data below shows the average monthly temperature for two communities over six months.

	April	May	June	July	Aug.	Sept.
Community A	17°C	19°C	21°C	23°C	25°C	23°C
Community B	15°C	17°C	19°C	21°C	23°C	21°C

   a) Write the rule for the pattern in the temperatures of Community A.
   b) Write the rule for the pattern in the temperatures of Community B.
   c) In a sentence, explain how the patterns are related.

4. One community received 2345 mm of rainfall in a year. Another community in a very dry part of the world received $\frac{1}{100}$ of that amount. Use two methods to calculate how much rainfall the second community received.

Chapter Review

5. a) Divide 871 by 10 and 100.

   b) Multiply 871 by 0.1 and 0.01.

   c) What relationship do you notice between dividing a number by 10 and 100, and multiplying the same number by 0.1 and 0.01?

6. Use the following formulas to find the perimeter and area of each rectangle.

   $P = 2l + 2w$

   $A = lw$

   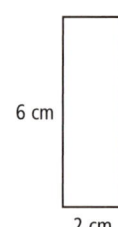
   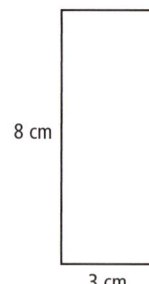

   6 cm, 2 cm, 8 cm, 3 cm

7. Find the missing factor "x" in each of the following:

   a) $9x = 63$ b) $56 = 7x$

   c) $8x = 72$ d) $256 = 16x$

8. Dan collected weather data on the temperature at his family's downtown home and compared it with the temperature at his grandparents' cottage north of the city. He found that the temperature at his home for this period was 4°C warmer.

   a) Write the relationship as an algebraic formula.

   b) Use the formula to find out what the temperature in the city was if the temperature at his grandparents' cottage was 28°C.

Chapter 3: Weather Patterns

9. **a)** On grid paper, draw a two-dimensional pattern of red and black squares.

   b) Use red and black cubes to turn this pattern into a three-dimensional pattern.

   c) Write a rule for your three-dimensional pattern.

10. Complete this factor table. Which of the factors on your chart is a variable? Which is a constant? How do you know?

Factor	2	3	4	5	6	7	8	9
Factor		8		8	8		8	
Product	16		32			56		81

Chapter Review 157

PATTERNING AND ALGEBRA

Chapter Wrap-Up

Create a Report on Local Weather Patterns

You have learned that there is a difference between weather and climate. Weather describes what is happening in the sky at one particular moment. Climate, on the other hand, describes the long-term weather patterns in a particular area.

Your mathematical learning in this chapter has focused on patterns and algebra. You have learned how to identify relationships between patterns, to recognize the relationship between the position of a number and its value, and to create patterns with two variables. You have also learned how to identify pattern rules, make predictions based on patterns in tables, and find missing factors.

Your task now is to work in a group of five to prepare a report on important weather patterns, or the climate, in your local area.

1. Meet with your group. Each group member must become an expert in one of the following climate topics:
   - sunshine
   - temperature
   - wind
   - air pressure
   - precipitation (for example, rainfall and snowfall)

   Decide on an overall project schedule and plan for completing the report on time. Help one another by sharing research sources and strategies.

2. Work on your own to research data about your topic. Identify important patterns in the data you collect. What are the rules for patterns? How are patterns related?

3. Find or create graphs, tables, and charts that clearly show important patterns related to your topic.

4. Meet in your groups to share, compare, and discuss your research findings. Decide on the best way to share your information so that the audience will obtain a good understanding of the local climate.

5. Make a weather prediction for an upcoming day, based on information you collected. At the end of this day, comment on the accuracy of your prediction. How reasonable was it?

6. Create missing-factor problems based on local weather patterns. Make sure you know the solution and can outline a clear, useful strategy for solving the problem.

7. Share your group's local climate presentation with the class. End the presentation by challenging members of your class to solve your pattern problems.

Tip

Publications and Web sites produced by Environment Canada are an excellent starting point for your climate research. Local newspapers also provide records of weather in your area. You might limit your climate study to a week, a month, or a year.

Journal

Make notes on what you learned about mathematical patterns from other groups' reports. Also make notes on what you learned about local climate and weather patterns from the presentations.

Chapter Wrap-Up 159

Problems to Solve

Here are some more fun problems for you to solve. You will be given a helpful problem-solving strategy for the first four problems. For the last problem, you get to choose a strategy to use.

Problem 6

Target Data

STRATEGY: MAKE AN ORGANIZED LIST

Putting together a list allows you to organize your thinking and review what you have done. You can identify important steps in a problem, or things you need to figure out to help you solve the problem.

OBJECTIVE:

Collect data and record results

Problem

You and your friend are throwing Velcro balls at a target like the one shown. The aim of the game is to get exactly 60 points in 3 tries. What are the different combinations of throws that could win the game? Use the "Make an Organized List" strategy to solve this problem.

Tip
Here is an example:
first throw = 60 points, second throw = 0 points, third throw = 0 points.

Chapter 3: Weather Patterns

Reflection

1. What did you know about the problem before you began to solve it?

2. What did you need to figure out?

3. How is your list organized?

4. Which rings are better not to get? Why?

5. Was the "Make an Organized List" strategy a good strategy to use? Explain why or why not.

6. What is the difference between the "Construct a Table" strategy and the "Make an Organized List" strategy?

7. Share your work with the class.

Extension

Change the problem so that the target has different numbers. Also change the final score to aim for. What are the different combinations that you can come up with for 3 throws to add up to the final score? What if a player only had 2 throws? Challenge a classmate to come up with 3 throws to add up to the final score on your target.

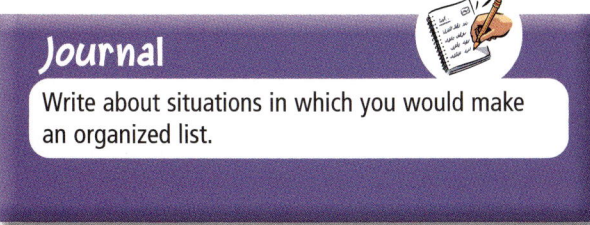

Journal
Write about situations in which you would make an organized list.

Problems to Solve 161

Problem 7
Building Measurements

STRATEGY: MAKE A MODEL

You can use different materials, such as egg cartons, construction paper, or craft sticks, to build a model for a problem. Using these materials helps you to see a solution to a problem.

OBJECTIVE:

Measure containers by volume using standard units: cubic centimetres

Problem

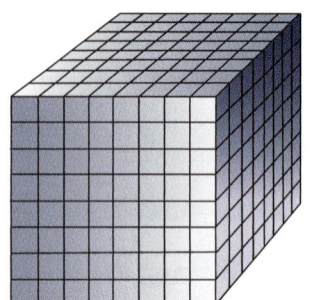

You Will Need
- linking cubes

Use the "Make a Model" strategy to help you solve these problems:

1. A building is 8 cubes long, 8 cubes wide, and 8 cubes high. If the building is solid, how many cubes do you need altogether?

2. Another building is 10 cubes long, 10 cubes wide, and 10 cubes high. How many cubes do you need to complete the solid building?

3. A third building is 12 cubes long, 12 cubes wide, and 12 cubes high. How many cubes do you need to make the building solid?

4. A fourth building is 14 cubes long, 14 cubes wide, and 14 cubes high. Complete the building to make it solid. How many cubes do you need for this building?

5. Is there a pattern in the total number of cubes needed? Explain the pattern.

6. Create a chart showing the pattern up to and including a building that is 20 cubes long, 20 cubes wide, and 20 cubes high.

Chapter 3: Weather Patterns

Reflection

1. What did you know about the problem before you started to solve it?

2. What did you need to figure out?

3. Do you think the "Make a Model" strategy was a good strategy to use? Explain why or why not.

4. Share your answers with another classmate.

Extension

With a partner, find an open-top box that you have permission to use. Estimate how many linking cubes could fill the box. Then, use linking cubes to find the actual answer. How close was your estimate to the actual volume of the box? Try this activity again using a different open-top box. Was your estimate more accurate this time?

Problem 8
Money Patterns

STRATEGY: LOOK FOR A PATTERN

Looking for a pattern is an important strategy that can be used for many kinds of problems. You can identify many different types of patterns, such as patterns in things you see or in the numbers you are working with. You can then continue the pattern to find the solution to a problem.

OBJECTIVE:

Identify and extend a pattern

Problem

After a big snowstorm, Dan decided to shovel his neighbours' driveways to earn some extra money. He shovelled 10 driveways altogether. The table below shows the money Dan earned for the first 4 driveways.

Assuming the pattern continued, how much money did Dan earn shovelling 5 driveways? 7 driveways? 10 driveways? Use the "Look for a Pattern" strategy to solve this problem and explain the pattern(s) you find.

Driveway	Money Earned
1	$3.00
2	$6.00
3	$9.00
4	$12.00

Reflection

1. What did you know about the problem before you began solving it?
2. What did you need to figure out?
3. How many different patterns did you find?
4. Do you think the "Look for a Pattern" strategy was a good strategy to use? Explain why or why not.

Extension

1. How much money did Dan earn in total? Remember he shovelled 10 driveways altogether. Use pictures, numbers, and words to support your answer.

2. If the pattern that you figured out continued, how much money would Dan have earned after shovelling 20 driveways? Show your work.

3. Create a problem that uses two or more number patterns. Give your problem to a classmate to solve.

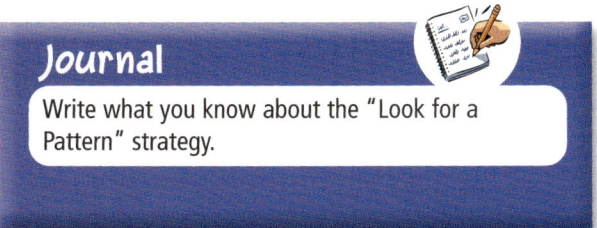

Journal

Write what you know about the "Look for a Pattern" strategy.

Problems to Solve

DATA MANAGEMENT AND PROBABILITY
PATTERNING AND ALGEBRA

Problem 9

Fish Ladder Patterns

STRATEGY: CONSTRUCT A TABLE

You can make a table to help you organize information, as well as spot missing information. Constructing a table makes looking for patterns easier, and it also highlights important information to help you solve the problem.

OBJECTIVE:

Discuss patterns in data

Problem

A fish ladder looks like a staircase. It is made so that fish can move up a river or canal that is dammed by a fence placed in the water. A biologist was doing a study of the fish that used a fish ladder. He observed that there were 100 fish on the first 5 steps of the ladder. In what pattern might the fish be arranged so that the biologist could predict the number of fish on the sixth, seventh, and eighth steps of the ladder? Use the "Construct a Table" strategy to solve this problem.

Reflection

1. What did you know about the problem before you started to solve it?

2. What did you need to figure out?

3. When you set up your table how many rows did you need?

4. Do you think the "Construct a Table" strategy was a good strategy to use? Explain why or why not.

5. Share and compare your chart with another classmate. Are your answers the same or different? If they are different, explain why.

Tip
Look back and read the question over again.

Extension

If you continued your pattern, how many fish would be on step 10?

Journal
Where have you seen tables of information used? Why do you think tables are used in these cases?

Problems to Solve

Unit 2
Math From the Past

Math has been used throughout history. In this unit, you will explore different cultures and study the amazing accomplishments that the knowledge of mathematics made possible.

In Chapter 4, you will learn about the designs and patterns of past civilizations, and about two-dimensional shapes, fractions, and decimals. In the Chapter Wrap-Up, you will apply your understanding of mathematics to create designs on a tile.

In Chapter 5, you will learn how measurement strategies were used by early peoples who designed and built great cities and by archeologists who study the past. In the Chapter Wrap-Up, you will apply your mathematical skills to create a site map that accurately describes sizes and shapes in an ancient community.

In Chapter 6, you will explore three-dimensional shapes such as pyramids, prisms, and cubes. In the Chapter Wrap-Up, you will use what you know to design, construct, and decorate a model of a monument.

Prepare for a journey of adventure and discovery!

Chapter 4

Patterns From the Past

People from early civilizations used lines, angles, and shapes to create art and decorate utensils and buildings. In this chapter, you will learn about some mathematical ideas that early peoples applied when creating designs. At the end of the chapter, you will apply these ideas to your own decorative tile design.

In this chapter, you will
- identify polygons by their sides and angles
- construct angles
- create congruent angles and shapes
- write equivalent fractions
- compare and order fractions and decimals to hundredths
- explore shape and number patterns and explain pattern rules

These questions will help you think about angles, fractions, and patterns:

1. Think about how fractions are used every day. Discuss your thoughts with a partner.
2. As a class, discuss the importance of angles in everyday life.
3. Where do you see patterns in everyday life?

Have fun exploring math from the past!

Chapter 4: Patterns From the Past   169

GEOMETRY AND SPATIAL SENSE

Lesson 1
Identifying and Describing Polygons

JOURNEY NOTE

PLAN:
You will identify and describe different shapes in a picture of an Incan wall. You will also construct various polygons.

DESCRIPTION:
The Inca people lived in South America's Andes mountain region from around A.D. 1000 to A.D. 1500. Incan cities were protected from enemies by walls made of huge stones of varied shapes.

The Incan city of Machu Picchu

Get Started

Here is a photograph of an ancient Incan wall showing how the differently shaped and sized stones fit together. Study the wall and describe the shapes you see to a partner.

170   Chapter 4: Patterns From the Past

Build Your Understanding

Describe Polygons

This Incan wall contains shapes that resemble polygons.

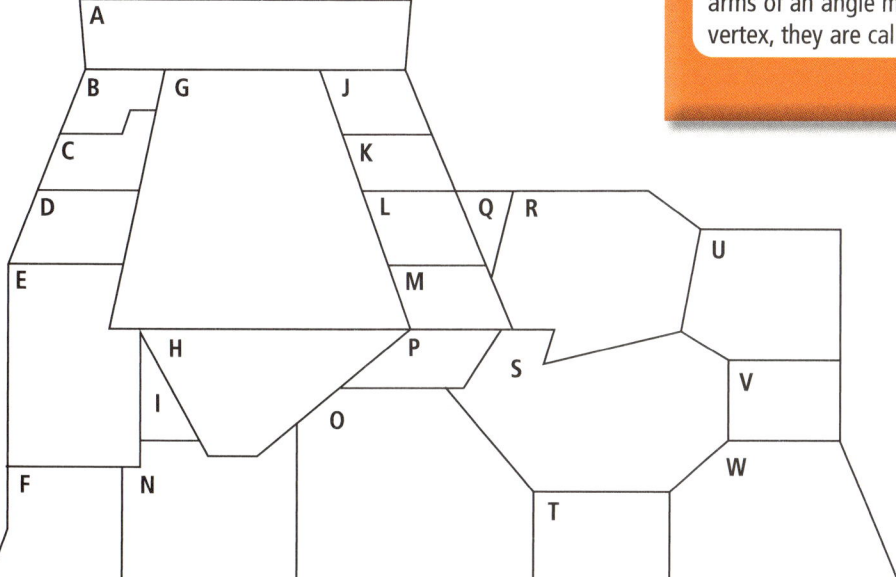

> **Vocabulary**
>
> **line segment:** A part of a line between two points on the line. Example: A———B
>
> **polygon:** A closed shape that is formed by three or more line segments. Some examples of polygons are triangles, quadrilaterals, and pentagons.
>
> **vertex:** The point where two line segments or arms of an angle meet. If there is more than one vertex, they are called vertices.

In a chart like the one below, complete the number of sides, angles, and vertices for each shape. If you know the name of the shape, enter it in the "Name" column.

Polygon	Name	Number of Sides	Number of Angles	Number of Vertices
A				
B				

> **Tip**
>
> If you can't remember or identify the names of all the polygons, look them up in the glossary, or share ideas with classmates. You might also revisit the chart after you complete the lesson, since the final activity involves naming polygons.

Lesson 1: Identifying and Describing Polygons

What Did You Learn?

1. In your notebook, explain how you decided one shape was different from another.
2. Describe the relationship among the number of sides, angles, and vertices of the various polygons.
3. Why do you think the Incas used such varied shapes and sizes in their walls? Why didn't they use rectangle-shaped bricks of equal size as do modern builders?

Practice

1. Draw the following polygons:
   a) A seven-sided shape.
   b) Three different shapes, each of which has four angles. Name your shapes.

2. Greek mathematicians from ancient times studied and named geometric shapes. From the Greek language, we have borrowed the following word parts:
   - *pente*, which means five
   - *hex*, which means six
   - *hepta*, which means seven
   - *octo*, which means eight
   - *deca*, which means ten

   From the Latin language, we have borrowed *nona*, which means nine.

   Use what you know about mathematics and the Greek and Latin roots of words to name all the polygons that you see in the design to the right. Share and compare with a classmate.

3. Draw three different polygons. Give them to a classmate to identify and describe. (Include the number of sides, angles, and vertices.) Share and discuss the results.

Technology

Use a works application (for example, AppleWorks, Microsoft Works) to create your own "Inca wall." Include various types of polygons. Use the spreadsheet tool to create a chart of sides, angles, and vertices. Paste the spreadsheet under your wall drawing.

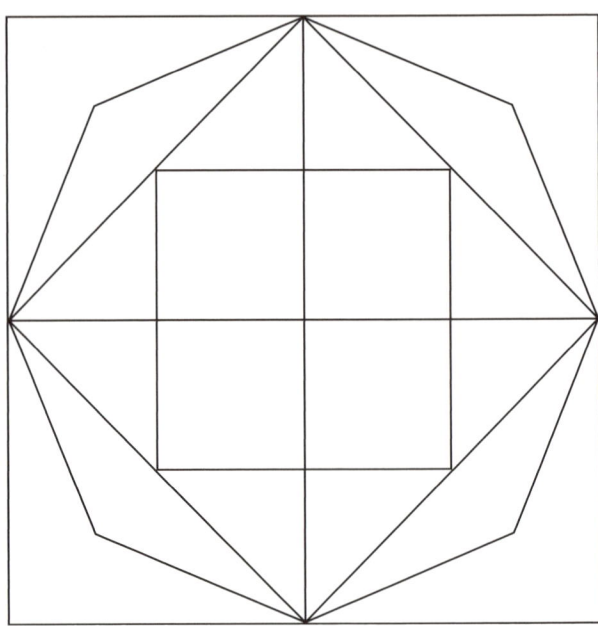

GEOMETRY AND SPATIAL SENSE
MEASUREMENT

Lesson 2

Classifying Triangles by Side Lengths

JOURNEY NOTE

An aryballos from Peru

An ancient Hawaiian petroglyph

Traditional hats

Pottery designs from early Greek civilization

PLAN:
You will classify and compare different types of triangles. You will also create various types of triangles using geoboards, pieces of drinking straws, or string.

DESCRIPTION:
Imagine you could visit a museum where the art of many early civilizations was on display. You might see an aryballos, a clay container from the ancient Incan civilization used to transport and store liquid. Incan pottery decorations often featured geometric designs, including triangles and diamond shapes.

Get Started

You Will Need
- geoboard
- elastic bands

Triangles appear in the designs of many cultures.

1. With a classmate, describe and compare the different triangles that you see in the photographs in the Journey Note.

Lesson 2: Classifying Triangles by Side Lengths

2. Find examples of isosceles, equilateral, and scalene triangles in the pictures or in your classroom. Explain to a classmate why the triangle is isosceles, equilateral, or scalene.

3. Using a geoboard and elastic bands, make a variety of triangles. Compare your triangles with those of a classmate. Describe the similarities and differences among the triangles. Use mathematics vocabulary.

Vocabulary

equilateral triangle: A triangle in which all sides are the same length
geoboard: A piece of wood or plastic that has rows of pegs or nails evenly spaced
isosceles triangle: A triangle that has two sides of equal length
scalene triangle: A triangle with three sides of different lengths

Tip

Consider using a ruler, a protractor, or other measurement tools as you explore and compare triangles.

Build Your Understanding

Construct Triangles With Different Side Lengths

You Will Need
- ruler
- drinking straws
- scissors

1. Can you make a triangle with sides of any length? Write a prediction in your notebook.

2. To test your prediction, you will work in a small group or on your own. Each group must cut three lengths of straw to be used as the sides of a triangle. If you are working alone, do this step yourself. Cut the lengths to the nearest whole centimetre.

Chapter 4: Patterns From the Past

3. Make a triangle using the three pieces of straw. Record the length of each side you used in a chart. If you can create a triangle with three sides of these lengths, write "Yes" in the fourth column. If you cannot, write "No." Record the type of triangle in the last column.

Length of Sides			Can a Triangle Be Formed?	Triangle Type
Side 1	Side 2	Side 3		

4. Share your results with other individuals or groups. Record their data and observations in your chart.

5. Exchange straws with other individuals or groups. Try to make triangles with other combinations of side lengths. Make sure you record all of your results, as well as the results of your classmates.

6. Revisit and comment on the prediction you made in question 1.

Journal

In your own words, write definitions of isosceles, equilateral, and scalene triangles. Use pictures with labels to help you understand and remember the characteristics of each type of triangle.

Technology

Create a computer file to record mathematical definitions. Revisit these definitions to make improvements to them based on new mathematical ideas you have learned. If possible, use computer programs to create pictures to support your definitions.

Lesson 2: Classifying Triangles by Side Lengths

What Did You Learn?

1. Make a large class chart to record all the individual and group triangle findings.

Triangle Type	Longest Side	Sum of Length of Other Two Sides

2. Discuss the relationship between the lengths of the sides and whether or not a triangle can be made.

3. Sort the triangles on your chart according to their side lengths. What patterns can you see?

4. In your notebook, describe the relationships you noted between the sides of a triangle.

5. Explain in pictures and words each of the following:
   a) an equilateral triangle
   b) an isosceles triangle
   c) a scalene triangle

6. Use a draw application (for example, AppleWorks, Corel Draw). Using the polygon tool and a grid background, draw and fill equilateral, isosceles, and scalene triangles. Create a legend to identify the triangles.

Practice

1. Predict what type of triangle each of the following figures is.

a)  b)  c)  d) 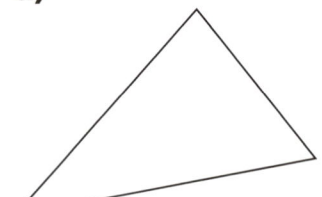 e)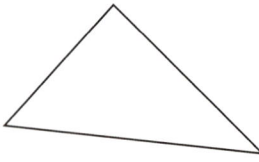

Chapter 4: Patterns From the Past

2. Check your prediction by measuring each triangle's sides. Were you right?

3. Draw a triangle of your choice. Describe it as fully as you can in your notebook. Read the description to a classmate. See if your classmate can draw your triangle by following the description. Do your triangles match? If not, how could you improve your description?

4. Find three different triangles in your classroom. Predict what type of triangle each one is. Check your prediction by measuring each triangle's sides. Were you right?

Extension

5. Predict which of the combinations of side lengths shown will create a triangle. Record your prediction in a chart by writing "Yes" or "No." Also, record a brief explanation for your prediction. Finally, use a geoboard, pieces of drinking straw, or string to test your predictions.

Length of Sides (cm)			Can a Triangle Be Formed?		
Side 1	Side 2	Side 3	Prediction	Explanation	Results
3	4	5			
10	5	4			
6	8	10			
9	12	6			
37	18	11			

6. In the chart below, the lengths of two triangle sides are provided. Predict what the length of the third triangle side might be. Give sound reasons for your predictions. Experiment with a geoboard, pieces of drinking straw, or string to test your predictions.

Triangle	Side 1 (cm)	Side 2 (cm)	Side 3 (cm)
A	4	6	
B		8	12
C	10		14

Lesson 2: Classifying Triangles by Side Lengths

Lesson 3
Classifying Triangles by Angles

JOURNEY NOTE

PLAN:
You will name and construct right, acute, and obtuse angles, and identify perpendicular lines. You will also name and construct right-angle, acute, and obtuse triangles.

DESCRIPTION:
If you look closely, you can see a variety of angles in the designs of many early cultures.

Nazca pottery

Near Eastern rug

Get Started

An angle is formed when two line segments meet. In the picture below, AB and BC meet at B. They form the angle ABC.

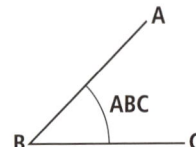

The segments used to form an angle are called "rays" or "arms." When angles are named, the order of the letters is important. The letter at the vertex, which is where the two line segments meet, must be the letter in the centre. When describing angles, use the angle symbol (∠). The angle pictured can be called ∠ABC or ∠CBA or ∠B (if no other angles exist at B).

Chapter 4: Patterns From the Past

∠DEF below is called a right angle. A right angle measures 90 degrees, or 90°. When two lines or line segments meet to create a right angle, we say they are perpendicular. In the picture, line segment DE is perpendicular to line segment EF.

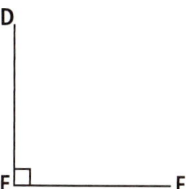

Vocabulary

acute angle: An angle that measures less than 90 degrees
angle: Line segments that share an endpoint. We measure angles in degrees (°).
obtuse angle: An angle that measures more than 90 degrees but less than 180 degrees
perpendicular: Meeting at a right angle, as lines or line segments do in the corner of a square
right angle: A square corner angle. A right angle is 90 degrees.

An acute angle is smaller than 90°.

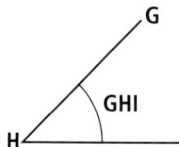

An obtuse angle is greater than 90° but smaller than 180°.

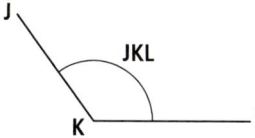

With a classmate, examine the photographs at the beginning of this lesson. Identify right, acute, and obtuse angles in the photographs. Discuss your findings with your partner.

Build Your Understanding

Classify Triangles by the Angles They Contain

You Will Need
• protractor

Triangles can be sorted according to the types of angles they contain.

Lesson 3: Classifying Triangles by Angles

Work with a partner.

1. For each triangle below, estimate the size of each angle it contains. Then use a protractor to measure each angle. Determine if the angle is right, acute, or obtuse.

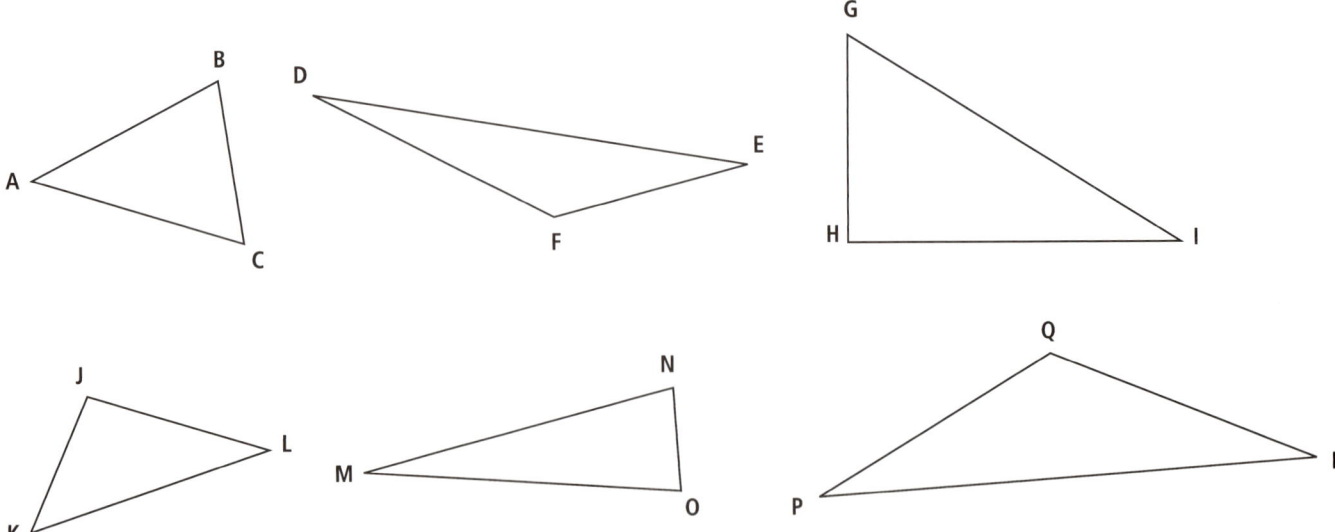

2. Record the number of each type of angle in a chart.

Triangle	Number of Right Angles	Number of Acute Angles	Number of Obtuse Angles
ABC			
DEF			
GHI			
JKL			
MNO			
PQR			

3. When the chart is complete, identify with your partner any of these types of triangles from the ones you explored:
   - any triangle that contains two obtuse angles
   - any triangle that contains an obtuse and two acute angles
   - any triangle that contains three acute angles

Chapter 4: Patterns From the Past

4. Read the shaded box. Go back to question 1 and name the triangles according to the angles they contain. Use the information in your chart to help you. Record your answers in your notebook. Discuss your findings with a partner.

Technology

If you are keeping a database of mathematical definitions, include your definitions for right-angle, acute, and obtuse triangles.

Triangles can be grouped in the following ways:
- A triangle that contains a 90° angle is called a right-angle triangle.
- A triangle in which each of the angles has a measure of less than 90° is called an acute triangle.
- A triangle that contains one obtuse angle is called an obtuse triangle.

What Did You Learn?

1. Draw, name, and label the following types of angles:
    a) right angle
    b) acute angle
    c) obtuse angle

2. Using pictures and words, describe each of the following types of triangles:
    a) right-angle triangle
    b) acute triangle
    c) obtuse triangle

3. What statements can you make about classifying triangles by the angles they contain?

4. Draw a triangle using a ruler. Determine if the triangle is right-angle, acute, or obtuse. Explain how you know.

Practice

1. Name the following triangles according to their angles. Explain your thinking.

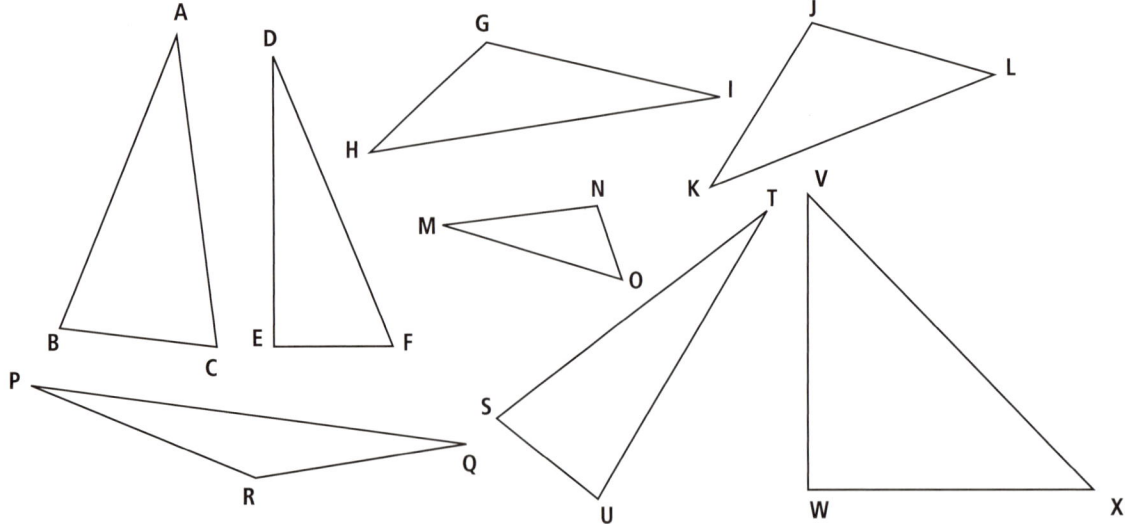

2. Use a draw application to create triangles using various types of angles. Have a partner name them according to their angles. After completion, discuss with your partner why you named each other's triangles the way you did.

Show What You Know
Review: Lessons 1 to 3, Identifying Polygons, Classifying Triangles

1. In your notebook, draw the following triangles:
   a) two different right-angle triangles
   b) two different obtuse triangles
   c) two different acute triangles

2. Apply what you know about triangles' side lengths and their angle measurements to create the following:
   a) triangle that is both right-angle and scalene
   b) triangle that is both obtuse and scalene
   c) triangle that is both obtuse and isosceles
   d) triangle that is both right-angle and equilateral

3. Find pictures of decorative designs from an early civilization that contain mathematical ideas discussed in this lesson. Explain to a small group of classmates how the mathematical ideas are represented in the picture.

Chapter 4: Patterns From the Past

GEOMETRY AND SPATIAL SENSE

Lesson 4
Identifying Congruent Shapes and Angles

JOURNEY NOTE

PLAN:
You will identify and create congruent shapes and angles.

DESCRIPTION:
Imagine you are part of a team of archeologists, people who study civilizations of the past. Your team uncovers a symbol from early China, where balance between sunlight (yang), and dark (yin) was important. Another team, in Ghana, uncovers pectoral discs that show a variety of congruent and repeating shapes.

Chinese yin yang symbol

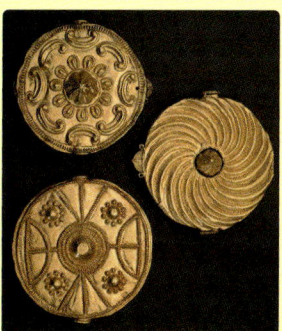
Pectoral discs from Ghana

Get Started

1. List the letters of any shapes that are exactly the same.

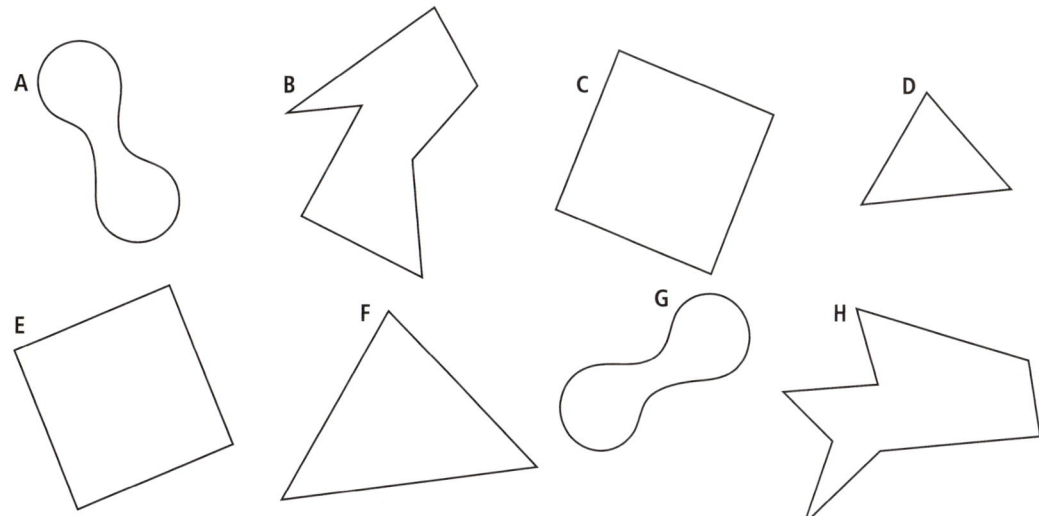

In what way are the shapes the same? How can you prove they are exactly the same?

Shapes that are exactly the same size and shape are congruent.

Lesson 4: Identifying Congruent Shapes and Angles

2. Use a Mira to find out if the shapes below are congruent.

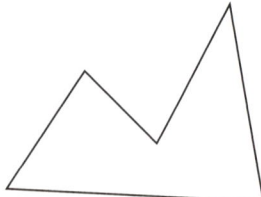 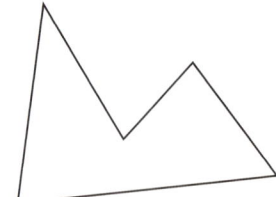

3. Use a geoboard or grid paper to construct at least three different sets of congruent shapes. Share your shapes with a classmate.

Tip
One way to determine if two shapes are the same is to trace one shape and see if it fits exactly over the second.

Vocabulary
congruent: Exactly the same size and shape

Angles can also be congruent. When angles are congruent, they are the same size, or measure exactly the same number of degrees. It does not necessarily mean the pairs of line segments that make up each angle have the same length. Which of these angles are congruent? Prove that they are congruent.

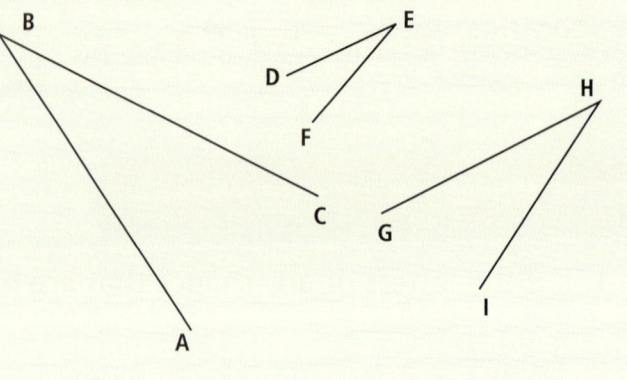

Journal
Make a list of strategies and tools you could use to prove that two or more shapes are exactly the same. Share your ideas with a classmate. Explain which strategies and tools you found most effective.

Booklink
Wonders of the World by Mark Bergin (Scholastic: Markham, ON, 1998). This book discusses the design and construction of the Seven Wonders of the Ancient World and compares them to impressive modern structures.

Chapter 4: Patterns From the Past

Build Your Understanding

Construct and Explore Congruent Shapes and Angles

1. Use tracing paper to create a shape that is congruent to the shape below.

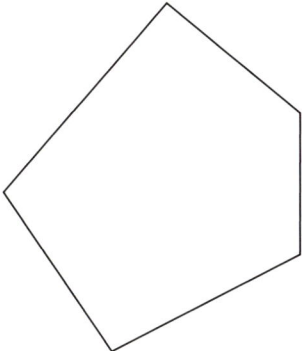

   What is the name of the congruent shape you created? Explain in your notebook the method you followed to create the congruent shape.

2. Use a method other than tracing to create angles that are equal to each of the angles below.

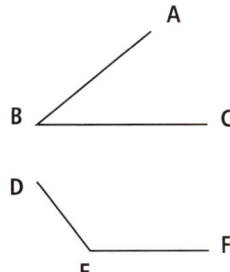

   Explain in writing the method you followed to create the congruent angles.

3. Make two congruent shapes of your own. How do you know they are congruent?

4. Construct three equal angles. Explain how you know all three angles are equal.

5. a) In two or more congruent shapes, are the respective angles in the shapes equal? Explain your thinking. Use pictures to help support your conclusion.

   b) Is just having equal angles enough to make two shapes congruent? Explain your thinking. Use pictures to help support your conclusion.

Lesson 4: Identifying Congruent Shapes and Angles

What Did You Learn?

1. Have a classmate check your answers to questions 1 to 4 of Build Your Understanding to determine if the shapes are congruent and the angles are equal.

2. Exchange your written answer to question 5 of Build Your Understanding with a classmate. Edit the classmate's work to make it clearer and more mathematically correct. When you return the edited work, explain why you suggested the changes you did.

3. What does congruent mean? Use pictures and words to explain your answer.

Tip

When editing a classmate's work, always point out the good parts of it first. Don't make general comments like "I didn't like it." Tell your classmate clearly what will make his or her work better.

Practice

1. Use the following shapes to answer the questions below.

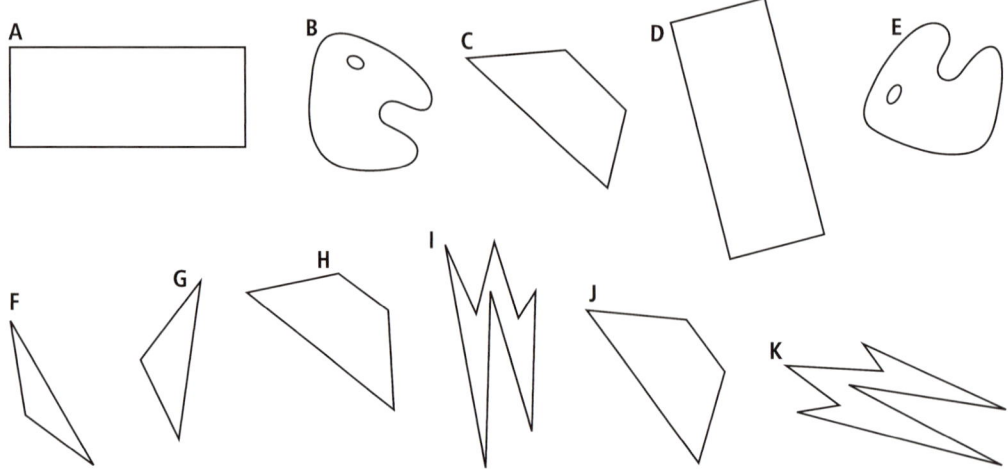

a) List the letters of any shapes that are congruent.
b) Explain why they are congruent.
c) Briefly describe the method(s) you used to determine whether shapes were congruent or not.

Tip

Make note of the materials, tools, and strategies you could use to create two congruent shapes.

Chapter 4: Patterns From the Past

2. Shape B below is incomplete. Use tracing paper to complete Shape B so that it is congruent with Shape A.

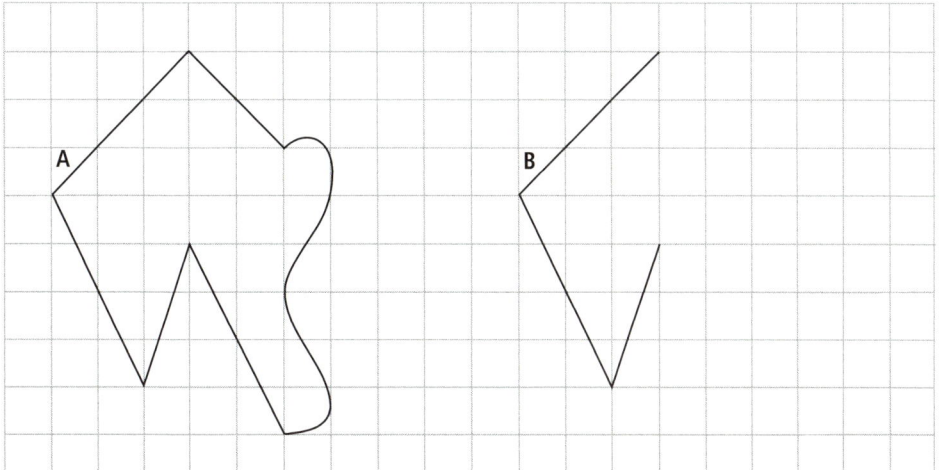

3. Draw a shape in which all the angles are equal.

4. Make a square, like the one below, in your notebook.

Technology
Use a draw application to explore congruent shapes. Use paint tools to create your own shapes. Reflect and rotate parts of the shapes to make sure they are congruent.

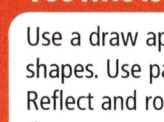

Within the square, find a way to create
a) two congruent shapes
b) four congruent shapes
c) eight congruent shapes

5. The three triangles below are equilateral triangles.

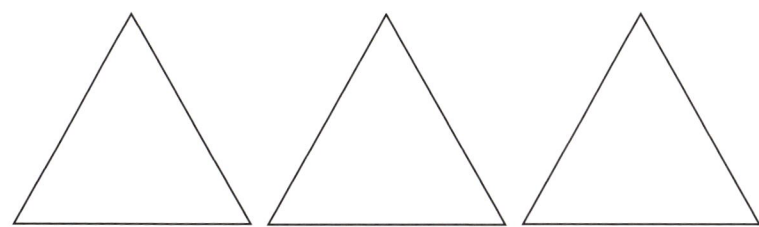

Make exact copies of these triangles in your notebook. Then, find a way to divide each equilateral triangle to create other congruent shapes.

Lesson 4: Identifying Congruent Shapes and Angles

NUMBER SENSE AND NUMERATION
PATTERNING AND ALGEBRA

Lesson 5

Identifying and Comparing Equivalent Fractions

JOURNEY NOTE

PLAN:
You will compare fractions and order fractions from the greatest to the least.

DESCRIPTION:
Many ancient civilizations used fractions. In the fifteenth century, a Persian mathematician named al-Khwrizmi wrote a book called <u>The Key to Arithmetic</u> for people, such as merchants, astronomers, and surveyors, who frequently used numbers. His book contained information about fractions.

This ancient Greek theatre shows the seating divided into equal sections.

Get Started

You Will Need
- ruler
- two rectangle reproducible pages, one showing sixteenths and the second showing twelfths

1. Work with a classmate. On the reproducible page containing the four rectangles with dashes, fill in the dotted lines to
   a) divide one rectangle into halves
   b) divide one into quarters
   c) divide one into eighths
   d) divide one into sixteenths

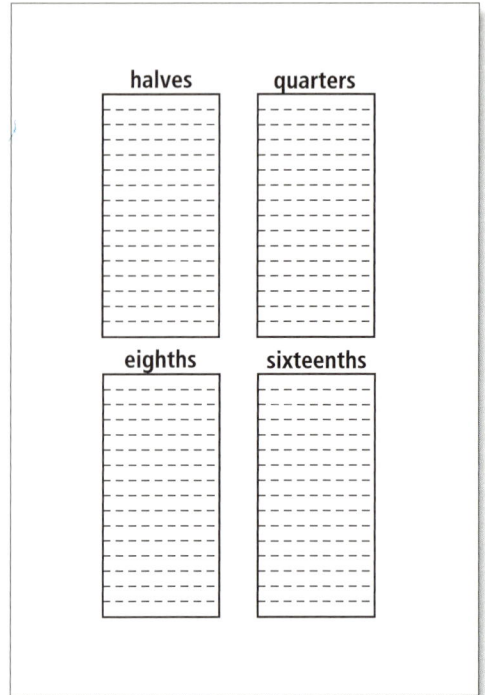

Chapter 4: Patterns From the Past

2. Look at the parts of the rectangles that represent the fractions $\frac{1}{2}$, $\frac{1}{4}$, $\frac{1}{8}$, and $\frac{1}{16}$. Put these fractions in order from the greatest to the least.

3. What happens to the denominator (the number on the bottom of each fraction) as the fractions get smaller? Using the rectangle pictures you created, explain why this makes sense.

4. Use the pictures you created to find how many quarters there are in a half. Write your answer as a statement.

   a) $\frac{1}{2} = \frac{\square}{4}$

   Next, use your rectangle pictures to complete the following:

   b) $\frac{1}{2} = \frac{\square}{4}$ or $\frac{\square}{8}$ or $\frac{\square}{16}$

   c) $\frac{1}{4} = \frac{\square}{8}$ or $\frac{\square}{12}$ or $\frac{\square}{16}$

   d) $\frac{1}{8} = \frac{\square}{16}$

5. On the reproducible page containing the three rectangles with dashes, fill in the dotted lines to

   a) divide one rectangle into thirds

   b) divide one into sixths

   c) divide one into twelfths

6. Look at the parts of the whole rectangles that represent the fractions $\frac{1}{3}$, $\frac{1}{6}$, and $\frac{1}{12}$. Put these fractions in order from the greatest to the least.

7. Use shading or colouring on your rectangles to find equivalent fractions for the following:

   a) $\frac{1}{3} = \frac{\square}{6}$ or $\frac{\square}{12}$
   b) $\frac{1}{6} = \frac{\square}{12}$
   c) $\frac{2}{3} = \frac{\square}{6}$ or $\frac{\square}{12}$

Journal
How did you know which fraction was greatest or least? Explain using pictures, numbers, and words.

Vocabulary
equivalent fractions: Fractions that represent the same part of a whole or a group. $\frac{1}{2}, \frac{2}{4}, \frac{3}{6}, \frac{4}{8}, \frac{5}{10}$, and $\frac{6}{12}$ are all equivalent fractions.

denominator: The bottom number in a fraction

$\frac{3}{4}$ ← denominator

numerator: The top number in a fraction

$\frac{1}{2}$ ← numerator

Tip
The numerator counts how many parts of a fraction are used. For example, a shape is divided into four equal parts with three of the quarters coloured. The fraction representing the total coloured parts is written $\frac{3}{4}$.

thirds	sixths	twelfths

Lesson 5: Identifying and Comparing Equivalent Fractions

Build Your Understanding

Divide Shapes to Show Equivalent Fractions

You Will Need
- reproducible page with different shapes
- reproducible page with sets of squares
- coloured pencils
- pattern blocks (optional)

1. On the reproducible page, divide the parallelogram (A) into halves and then into quarters. What equivalent fractions do you see? Record your answers next to the parallelogram.

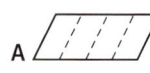

2. Divide the hexagon (B) into halves and then into sixths. What equivalent fractions do you see? Record your answers next to the hexagon.

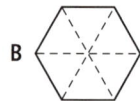

3. Divide the shape labelled C into quarters and then into eighths. What equivalent fractions do you see? Record your answers next to the shape.

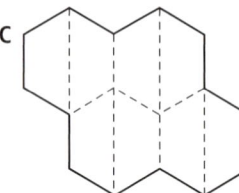

4. Divide the connected shapes (D) into thirds and then into sixths. What equivalent fractions do you see? Record your answers next to the shape.

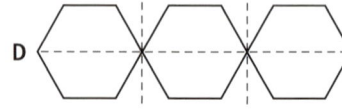

5. Divide the parallelogram (E) into fifths and then tenths. What equivalent fractions do you see? Record your answers next to the parallelogram.

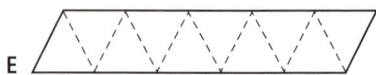

6. Look at the groups of squares in Set A and Set B.
   a) What equivalent fractions do you see in Set A?
   b) Set B could stand for "1." Using the diagram, write an equivalent fraction for "1."
   Record your answers next to each set.

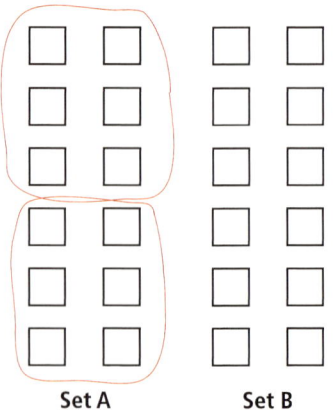

Chapter 4: Patterns From the Past

What Did You Learn?

1. With a classmate, compare and discuss your answers for questions 1 through 6 of Build Your Understanding. Add any equivalent fractions you didn't get to your sheets.

2. On a piece of scrap paper, create your own shape. Divide the shape into parts to show equivalent fractions. Exchange your drawing with a partner, and challenge him or her to identify as many equivalent fractions as possible.

Practice

1. Order the following fractions from the greatest to the least:

   $\frac{1}{8}$ $\frac{1}{16}$ $\frac{1}{4}$ $\frac{1}{2}$ $\frac{1}{3}$ $\frac{1}{6}$

2. Write the following pairs of fractions in your notebook, then circle the larger fraction in the pair. Explain how you know you circled the larger fraction.

   a) $\frac{1}{4}$, $\frac{3}{4}$ b) $\frac{1}{5}$, $\frac{1}{6}$

   c) $\frac{1}{2}$, $\frac{2}{3}$ d) $\frac{3}{8}$, $\frac{1}{2}$

Tip

If you have difficulty deciding which fraction is larger or smaller, you might divide rectangles as you did earlier in the lesson.

Technology

Use a spreadsheet application with square cell formatting. Make and label equivalent fractions by filling cells with colour. Use the text tool to label your cells, and highlight cell borders to outline your fractions.

Math Problems to Solve

3. Kenny and Shefali divided one apple. Shefali ate $\frac{3}{4}$ of the apple. How much was left for Kenny?

4. Jeffrey and Aisha each have the same size pizza for lunch. Jeffrey ate $\frac{4}{5}$ of his pizza. Aisha's pizza has 10 slices in it. If Aisha wants to eat the same amount of pizza as Jeffrey, how many slices will she have to eat? Use pictures, numbers, and words to show your answer.

Lesson 6

Using Multiplication to Create Equivalent Fractions

JOURNEY NOTE

PLAN:
You will review what you know about equivalent fractions. Then you will explore what happens when you multiply the numerator and denominator of a fraction by the same number.

Colosseum, Rome, Italy

DESCRIPTION:
Imagine you could attend a major sporting event in Rome during the days of the Roman Empire. You would likely watch the event at the Colosseum, a stadium that held about 50 000 people and had six circular levels. About $\frac{1}{6}$ of the spectators sat on each level of the stadium.

Get Started

1. How many $\frac{1}{8}$ are there in $\frac{2}{4}$? Name another fraction that is equivalent to $\frac{2}{4}$. How did you get your answer?

2. Which fraction is larger, $\frac{7}{8}$ or $\frac{3}{4}$? How can you tell?

Tip

Study the fraction strip charts on the next page with a classmate, and explain how to read each chart. You might test your classmate's understanding by creating questions for him or her. For example, how many $\frac{1}{4}$ are there in $\frac{1}{2}$?

Chapter 4: Patterns From the Past

3. With a partner, create other questions that could be answered by using the fraction strips. Share and discuss your questions with the class.

Fraction Strip Charts

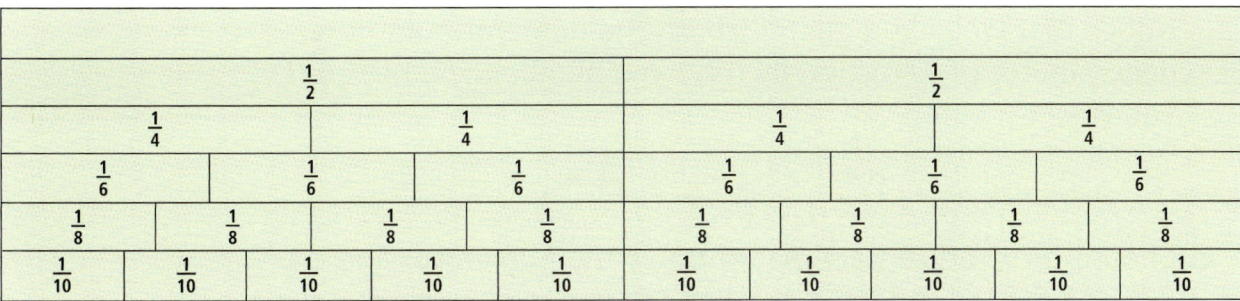

Fractions you explored, such as $\frac{1}{2}$, $\frac{2}{4}$, and $\frac{4}{8}$, are equivalent fractions. All of these fractions represent the same part of a whole or group.

Build Your Understanding

Explore Numerators and Denominators in Equivalent Fractions

1. With a partner, use the fraction strip charts to make a list of fractions that are equivalent to the following fractions. Make your lists as long as possible.

   a) $\frac{1}{2} = $ ■, ■, ■, ...   b) $\frac{1}{3} = $ ■, ■, ■, ...   c) $\frac{1}{4} = $ ■, ■, ■, ...

> **Journal**
>
> Compare the lists you made. Make note of any relationships or patterns you notice in the numerators and denominators of equivalent fractions.

Lesson 6: Using Multiplication to Create Equivalent Fractions

2. With a classmate, choose one set of equivalent fractions from question 1. What has been done to both the numerator and denominator to create an equivalent fraction? Write a sentence to summarize your partner's and your conclusion.

Notice that when you multiply both the numerator and the denominator of a fraction by the same number, you get an equivalent fraction.

$$\frac{1}{2} \times \frac{2}{2} = \frac{2}{4} \qquad \frac{1}{2} \times \frac{3}{3} = \frac{3}{6} \qquad \frac{1}{2} \times \frac{4}{4} = \frac{4}{8}$$

3. Choose three other fractions. Multiply the numerator and denominator by the same number. Did you create an equivalent fraction in each case? Write your conclusion in a sentence. Use pictures or fraction strips to help support your conclusion.

Tip

If you have trouble understanding equivalent fractions, you might create and label sets of fraction strips like the ones in the charts on the previous page. Use your fraction strips to explore the relationships among fractions.

What Did You Learn?

1. Explain what you have learned about the relationship between multiplication and the numerator and denominator of fractions. Use fraction strips or pictures, numbers, and words to explain your ideas. Share and discuss your explanation with a classmate.

2. List at least five equivalent fractions for each fraction below.

   a) $\frac{1}{2}$ b) $\frac{1}{3}$ c) $\frac{3}{4}$ d) $\frac{4}{5}$

   Explain to a classmate how you created equivalent fractions for each fraction.

3. Are the following fractions equivalent?

   $\frac{7}{12} \qquad \frac{16}{24}$

   How do you know? Meet with a classmate to share and discuss your conclusions.

Chapter 4: Patterns From the Past

Practice

Name at least two equivalent fractions for each of the following fractions:

1. $\frac{6}{10}$ **2.** $\frac{10}{15}$ **3.** $\frac{3}{6}$ **4.** $\frac{12}{15}$

Use the fraction strip charts in Get Started to decide if the first fraction is equal to (=), greater than (>), or less than (<) the second one. Copy each question into your notebook. Fill in the box with the correct symbol.

5. $\frac{3}{4}$ ☐ $\frac{2}{4}$ **6.** $\frac{3}{5}$ ☐ $\frac{4}{6}$ **7.** $\frac{2}{8}$ ☐ $\frac{4}{7}$

8. On a number line, numbers are arranged from least to greatest.

Make a number line for all the fractions in question 2 of What Did You Learn? Order the fractions on your number line from least to greatest. Have a classmate check your work.

9. Make a list of six fractions. Have a classmate order the fractions on a number line from least to greatest. Share and discuss the results.

Vocabulary

= is equal to
< is less than
> is greater than

Technology

Use a calculator to demonstrate your understanding of equivalent fractions. Working in pairs, have one partner identify a fraction while the other uses a calculator to give two equivalent fractions. The first partner can check for accuracy using his or her calculator.

A Math Problem to Solve

10. Omar, Carla, and Avi each have the same size chocolate bar. Omar ate $\frac{2}{4}$ of his chocolate bar. Carla ate $\frac{6}{8}$ of her chocolate bar, and Avi ate $\frac{5}{6}$ of his chocolate bar. Who ate the most chocolate? Use pictures, numbers, and words to support your answer.

NUMBER SENSE AND NUMERATION

Lesson 7

Exploring Fractions and Decimals to Tenths

JOURNEY NOTE

PLAN:
You will explore the relationship among fractions, mixed numbers, and decimals using drawings. You will then construct a number line to help you write fractions and mixed numbers as decimal numbers. Finally, you will use the number line to compare and order decimal numbers.

DESCRIPTION:
Many of the incredible buildings left by the ancient Maya, like the famous Temple of the Jaguars at Chichén Itzá, Mexico, could not have been built if the Maya were not skilled in mathematics. The Mayan number system was similar to our own decimal system.

Mayan ruins at Tulum, Mexico

Get Started

You Will Need
- red and blue coloured pencils
- ten-square rectangle

1. Trace onto a piece of paper the rectangle shown to the right.

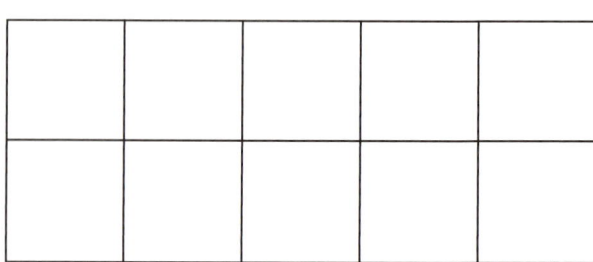

2. Colour some of the squares within the rectangle red and the remaining squares blue.

   Write a fraction for the part of the whole rectangle that is red. Write a fraction for the part of the whole rectangle that is blue.

Chapter 4: Patterns From the Past

3. Share, compare, and discuss your drawing and fractions with a classmate.

Fractions can also be written as decimal numbers.

As a decimal number, $\frac{1}{10}$ is 0.1. What would $\frac{2}{10}$ be as a decimal?

> **Vocabulary**
> **decimal number:** A number written in the decimal system. For example, 0.5, 0.6, or 1.3.
> **mixed number:** A whole number and a fraction together. For example, $2\frac{1}{3}$ is a mixed number.

4. Write the fractions you wrote to describe your red-and-blue rectangle as decimals.

5. Make a chart like the one below in your notebook, and then complete it to $\frac{10}{10}$.

Fraction	Decimal	Words
$\frac{1}{10}$	0.1	one tenth
$\frac{2}{10}$		
$\frac{3}{10}$		

6. What is the fraction when all of the sections are coloured blue?

7. Mixed numbers can also be written as decimals.

Write mixed numbers and then decimals to describe the shaded portions of the shapes below.

a)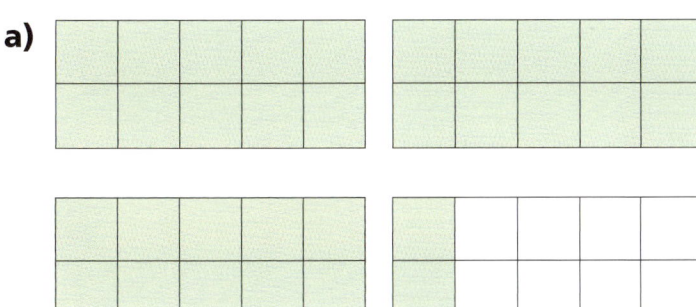

b)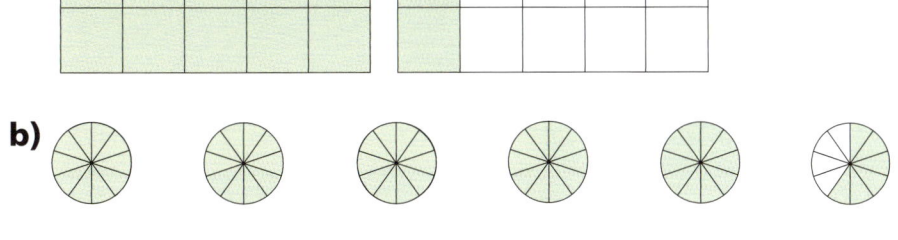

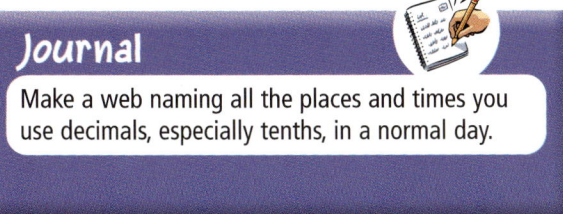

Journal
Make a web naming all the places and times you use decimals, especially tenths, in a normal day.

Lesson 7: Exploring Fractions and Decimals to Tenths

Build Your Understanding

Construct and Use a Tenths Number Line

You Will Need
- centimetre ruler
- 1-cm grid paper

1. Draw a long horizontal line on the paper you have been given.

2. Mark off five units on the line. Use the 1 cm by 1 cm squares on your grid paper as guidelines for your marks.

3. Mark off tenths in each unit.

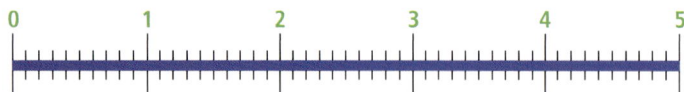

4. Mark and write the following mixed numbers on your number line:

   $1\frac{2}{10}$ $2\frac{3}{10}$ $4\frac{3}{10}$

5. Beneath each fraction, write the decimal equivalent.

6. Name fractions and decimals with a classmate, and find them on the number line. Check to see if the placement of each fraction or decimal is correct.

What Did You Learn?

1. Use pictures, numbers, and words to explain what 0.1 is equal to.

2. Put the following numbers in order from the least to the greatest:

   0.7   6.8   4.4   10.0   8.1   16.7

   How do you know your order is correct?

Tip

Discuss with a classmate how knowing about place value can help you to order numbers quickly and correctly.

Practice

Write each of the following fractions or mixed numbers as a decimal:

1. $\frac{2}{10}$
2. $\frac{4}{10}$
3. $1\frac{2}{10}$
4. $17\frac{6}{10}$

Write each of the following as a decimal. For example, $\frac{3}{10}$ as a decimal is 0.3.

5. $5\frac{4}{10}$
6. $6\frac{2}{10}$
7. $2\frac{6}{10}$
8. $14\frac{3}{10}$
9. $10\frac{1}{10}$
10. $40 + 7 + 0.9$

11. $900 + 6 + 0.2$

12. 6 tens, 4 ones, and 4 tenths

13. 8 hundreds, 7 tens, and 2 tenths

Make a drawing to show each of the following:

14. 3.4
15. $4\frac{1}{10}$
16. 7.6
17. $2\frac{9}{10}$

Journal

Read the weather section in a newspaper for a week. Make notes on how decimal numbers are used to give information. Share and compare information you collected with the class.

Tip

Here is a place-value reminder to help you:

2 3 6 . 5

hundreds   tens   ones   tenths

Copy each question into your notebook. Fill in the box with <, >, or =.

18. $3\frac{3}{10}$ ☐ 6.2
19. 1.7 ☐ $1\frac{5}{10}$
20. $\frac{9}{10}$ ☐ 0.9
21. 3 ☐ 3.1

22. Write the following fractions, mixed numbers, and decimals on a number line:

$4\frac{9}{10}$ 7.2 $\frac{2}{10}$ 2.9 $8\frac{8}{10}$ 1.1   9.9 $5\frac{5}{10}$

23. Make a list of six decimals like ones you looked at in this lesson. Give your list to a partner to order from least to greatest. Share and discuss the results.

Lesson 8

Exploring Fractions and Decimals to Hundredths

JOURNEY NOTE

PLAN:
You will use concrete materials, grid drawings, and numbers to explore the relationships between fractions and decimals to the hundredths place. Then you will practise reading, writing, comparing, and ordering decimals to hundredths.

DESCRIPTION:
If you lived in an early civilization and you needed to buy something, you might have used a form of money. Today, money amounts are often written as decimals.

An early Byzantine coin

Get Started

This grid is divided into 100 equal parts. Each small square part is $\frac{1}{100}$ of the whole. The fraction $\frac{1}{100}$ can be expressed as the decimal 0.01. You will notice that 47 parts out of the 100 squares are shaded. That means $\frac{47}{100}$ of the whole grid is shaded. Or, as a decimal number, 0.47 of the whole grid is shaded.

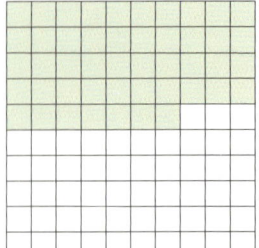

In Lesson 7, you learned about decimal numbers to the tenths place. In this lesson, you will learn about decimal numbers to the hundredths place.

You Will Need
- square of grid paper, 10 units by 10 units
- five different coloured pencils

1. Make a chart like the one below in your notebook.

Colour Used	Fraction	Decimal

2. On your grid paper, use five different colours to create an interesting pattern. Make sure you colour complete squares within the grid.

3. Record information about the pattern you created on the chart. Let's use the first row as an example. If you used the colour red, in the first column of the chart colour the first space red. In the second column, write a fraction to represent the number of squares that are red out of the total number of squares. In the third column, express that fraction as a decimal number.

4. When you complete your chart, exchange it and your coloured grid pattern with a classmate. Check each other's work.

5. With your classmate, discuss how the ideas you explored in this activity can be related to Canadian money.

Lesson 8: Exploring Fractions and Decimals to Hundredths

Build Your Understanding

Explore Decimal Numbers to Hundredths

1. For each grid, give the fraction and the decimal that tells what part is shaded.

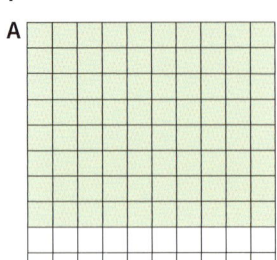
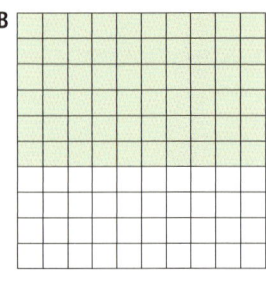
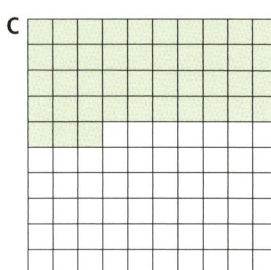

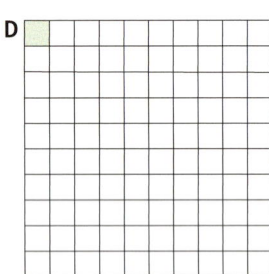
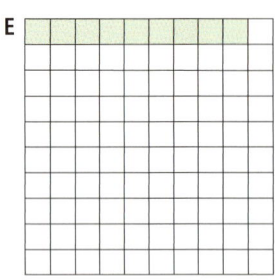
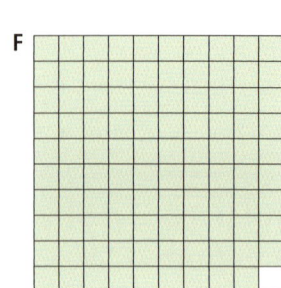

2. Write a list of the decimal numbers from question 1. Then, order the decimal numbers from least to greatest. Look at the grids above for help.

   How do hundredths relate to tenths?

   $\frac{1}{10} = \frac{10}{100}$

   Therefore, 0.1 = 0.10.

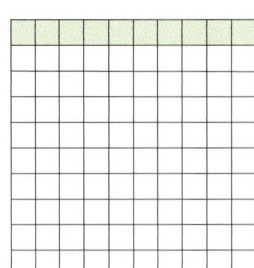

Journal

What relationship do you notice between tenths and hundredths? Use pictures, numbers, and words to explain your answer.

3. Use what you know about the relationship between tenths and hundredths to complete the following questions:

   a) $\frac{3}{10} = \frac{\blacksquare}{100}$ 0.3 =

   b) $\frac{6}{10} = \frac{\blacksquare}{100}$ 0.6 =

Tip

You can draw and shade grids like the example in question 2 to better understand the relationship between tenths and hundredths. Or you can use blocks, if they are available in your classroom.

Chapter 4: Patterns From the Past

What Did You Learn?

1. With a classmate, write a statement about the relationship between tenths and hundredths. Share and compare your statement with those written by other pairs. If necessary, revise your statement based on what you learned from others.

2. Write each of the following fractions as decimals:
   a) $\frac{23}{100}$ b) $\frac{79}{100}$ c) $\frac{34}{100}$ d) $\frac{8}{100}$

3. Complete the following question: $\frac{4}{10} = \frac{\blacksquare}{100}$ $0.4 = \blacksquare$

   Draw and shade grids (like the example in question 2 of Build Your Understanding) to support your answer.

4. Ask an adult you know if, and how, he or she uses decimal numbers, including hundredths, in his or her work. Share and discuss your findings with the class.

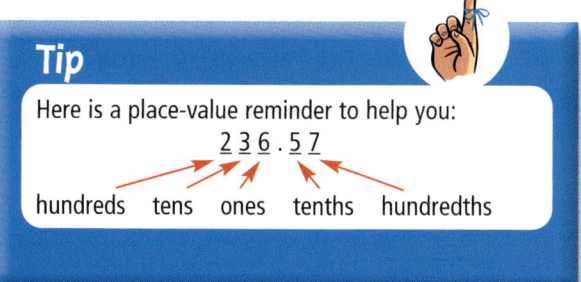

Tip
Here is a place-value reminder to help you:
2 3 6 . 5 7
hundreds tens ones tenths hundredths

Practice

Create shaded grids to show each of the following, then write each as a decimal.

1. 75 hundredths
2. $\frac{3}{10}$
3. 67 hundredths
4. $\frac{99}{100}$
5. $\frac{3}{100}$
6. $\frac{7}{10}$

7. Can the fraction $\frac{1}{2}$ be expressed as a decimal? If so, how?

Write these fractions as decimals:

8. $\frac{1}{4}$ 9. $\frac{3}{4}$ 10. $\frac{2}{5}$

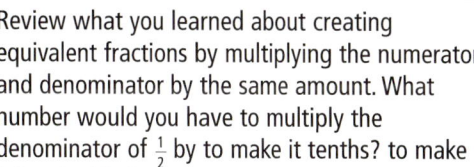

Tip
Review what you learned about creating equivalent fractions by multiplying the numerator and denominator by the same amount. What number would you have to multiply the denominator of $\frac{1}{2}$ by to make it tenths? to make it hundredths?

Journal
Write a list of simple steps to follow when changing a fraction into a decimal.

Lesson 8: Exploring Fractions and Decimals to Hundredths

Show What You Know

Review: Lessons 4 to 8, Identifying Congruent Shapes and Angles, Exploring Fractions and Decimals

1. Explain what equivalent fractions are using pictures, numbers, and words.

2. How can you use multiplication to create equivalent fractions?

3. Using materials and ideas from the past lessons, demonstrate to a small group of classmates why $\frac{3}{6}$ is the same as $\frac{1}{2}$.

4. Prove that 0.5 is equal to 0.50. Use pictures, numbers, and words.

5. Using materials and ideas from past lessons, demonstrate to a small group of classmates why $\frac{7}{10}$ is the same as 0.7.

6. Where would you or your family use fractions and decimals outside of school? Make a list and share it with a partner.

PATTERNING AND ALGEBRA

Lesson 9
Repeating and Expanding Patterns

JOURNEY NOTE

PLAN:
In this lesson, you will learn what repeating and expanding patterns are. You will use pattern blocks to create and extend repeating patterns and expanding patterns.

DESCRIPTION:
Lines, angles, and shapes have been used for thousands of years to create beautiful patterns. Study the picture here. Describe the patterns you see.

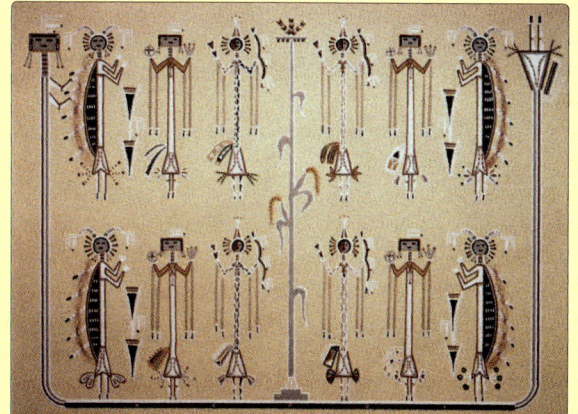

This Navajo sand painting shows a repeating pattern.

Get Started

1. Study these patterns. What figures create each pattern? How are all of the patterns similar? How are they different? How would you extend each pattern?

   The patterns are called repeating patterns. In each, a pattern is established and then repeated. For instance, in the first example, the square-circle pattern is repeated.

Journal

Note where you have seen interesting patterns in the world around you. You can sketch these patterns, showing their repeating parts. Write a definition of "pattern" based on what you know. Revisit your definition during and after the lesson, and, if necessary, revise the definition.

Lesson 9: Repeating and Expanding Patterns 205

2. Study these patterns. How are the patterns similar? How are they different?

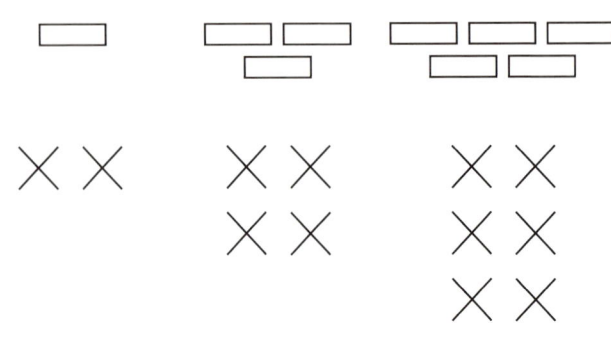

The patterns are called expanding patterns because with each extension of the pattern, new elements are added. For example, the first expanding pattern starts with one rectangle, in the next element of the pattern there are three rectangles, and in the third element there are five rectangles. What will be the next element in the expanding pattern?

Vocabulary

expanding pattern: A pattern in which the elements of the pattern increase with each extension

repeating pattern: A pattern in which the same elements repeat

Build Your Understanding

Create Repeating and Expanding Patterns

You Will Need
- two sets of pattern blocks

Work with a partner.

1. Place a barrier between you and your partner so your partner cannot see what you are doing.

2. Create a repeating pattern with a set of pattern blocks.

3. Describe the pattern to your partner using mathematical language. Be as clear and accurate as possible, so your partner can create a repeating pattern that is exactly the same as yours.

Tip

Write your pattern directions as clear step-by-step instructions using accurate mathematical language. Review your notes and the glossary at the end of this book to make sure you use the most accurate mathematical words possible.

4. Look at your partner's pattern. How similar is it to yours? How is it different? Is it a repeating pattern? How can you tell?

5. Switch roles.

6. Repeat steps 1 to 5 to create an expanding pattern. Explain why the pattern you created is an expanding pattern and not a repeating pattern.

Chapter 4: Patterns From the Past

What Did You Learn?

1. Share your patterns with a small group of classmates. Ask them to describe each of the patterns you created. Then, challenge them to indicate which pattern is repeating and which is expanding. Classmates must give reasons to support their conclusions.

2. Using your own words, explain what the difference is between a repeating pattern and an expanding pattern. Use pictures to help support your answer.

Practice

Extension

Copy each of these patterns into your notebook. Then, extend each of the patterns. Explain to a classmate why you extended the pattern the way you did.

1.

2.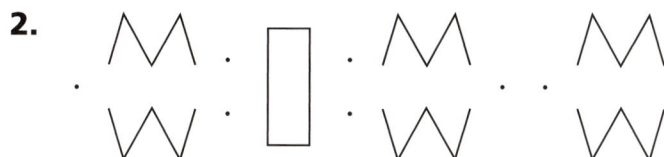

3.

4.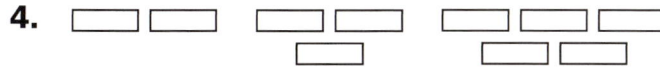

PATTERNING AND ALGEBRA
NUMBER SENSE AND NUMERATION

Lesson 10

Identifying, Extending, and Creating Number Patterns

JOURNEY NOTE

PLAN:
You will identify, extend, and create your own number patterns, as well as explain number pattern rules.

DESCRIPTION:
The early people of Babylon lived in what is now the Middle East between 1900 and 1300 B.C. They were great traders and among the first people to use money. Babylonians left many accomplishments; they were interested in numbers and built the Hanging Gardens of Babylon — one of the Seven Wonders of the Ancient World!

The Hanging Gardens of Babylon

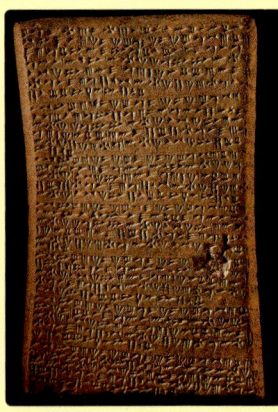
The Babylonians used a type of picture writing called cuneiform, which they wrote on clay tablets.

Get Started

In Lesson 9, you learned that different types of patterns can be made with lines, angles, and shapes. Patterns can also be made with numbers.

1. Look at the number patterns below.

   What are the next three numbers in each sequence? How do you know?

   a) 1, 5, 9, 13, …

   b) 111, 105, 99, 93, …

2. Describe the pattern rule for the patterns in 1 (a) and (b).

3. Create two number patterns to share with a classmate. Use a piece of scrap paper.

Tip
When trying to identify number patterns, ask yourself: What had to be done to the first number to get to the second number in the pattern? What had to be done to the second to get to the third? Continue asking these questions until you identify the pattern rule.

Vocabulary
sequence: A succession of things, including numbers, that are connected in some way; for example, the sequence of numbers 1, 2, 2, 3, 3, 3, 4, 4, 4, 4, …

Chapter 4: Patterns From the Past

4. On the back of the scrap paper, write the rule for continuing the pattern.

5. Challenge a classmate to identify each pattern rule. Apply the same pattern rule to create a new sequence with a different starting number.

> Number patterns fascinated people from early civilizations. On one clay tablet from around 1800 B.C., the following number pattern was found: 1, 4, 9, 16, 25, 36, 49, …
>
>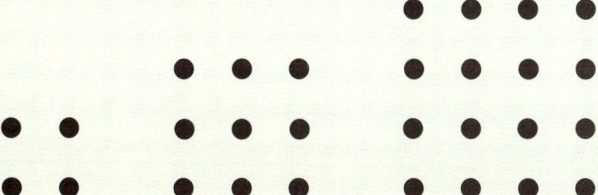
>
> This number pattern has become known as "the sequence of squares" because the numbers can be represented by dots arranged in squares.

6. What are the next three numbers in the pattern? How do you know?

7. What is the pattern rule for "the sequence of squares"?

8. What is the tenth number in this sequence? How do you know? Share and discuss your explanation with the class.

Journal

Give three examples of a sequence of numbers. Then, in your own words, write a simple, clear definition of "sequence." Check your definition of "sequence" in a general dictionary and in a mathematical dictionary.

Tip

If you are unsure of the meaning of any mathematical word, check it in a mathematical glossary, dictionary, or encyclopedia. You might keep a personal list, perhaps a computer database, of mathematical words and definitions that give you particular difficulty.

Build Your Understanding

Identify and Create Number Patterns

The sequence of squares is not the only sequence that can be shown as dots arranged in a geometric shape. This set of numbers is called "the sequence of triangles": 1, 3, 6, 10, …

1. Use dots to show why the pattern is called "the sequence of triangles."

2. What are the next three numbers in this sequence? What is the pattern rule?

3. What is the tenth number in this sequence? How do you know?

What Did You Learn?

1. Using pictures and words, explain what the sequence of a square pattern is. Share and discuss your explanation with a classmate.

2. Using pictures and words, explain what the sequence of a triangular pattern is. Share and discuss your explanations with a classmate.

3. Create four different patterns. Each pattern's rule must use a different mathematical operation (addition, subtraction, multiplication, or division). Challenge a classmate to identify each pattern rule.

4. What strategies do you use when trying to identify number patterns? Use examples to help support your answer.

5. What is the most important thing you learned about number patterns from discussing ideas with a classmate? Write down your explanation.

Practice

Give the next three numbers for each of these number patterns. Then, write a sentence explaining each pattern rule.

1. 91, 100, 109, 118, ■, ■, ■
2. 125, 235, 345, 455, ■, ■, ■
3. 346, 323, 300, 277, ■, ■, ■
4. 2, 5, 4, 7, 6, ■, ■, ■
5. 8, 12, 16, 20, ■, ■, ■

Extension

6. Here is another number sequence found on ancient Babylonian clay tablets: 1, 3, 9, 27, 81, ...

   a) What are the next three numbers in the pattern?
   b) What is the pattern rule?
   c) Explain how you figured out the pattern rule.

Technology

How might a calculator help you unlock the pattern? Use a calculator to identify the number pattern.

7. In 1653, a French mathematician named Blaise Pascal described a triangular arrangement of numbers that has come to be known as Pascal's Triangle. Here is the triangle:

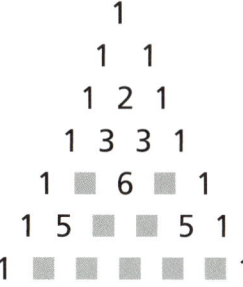

Booklink

Discovering Patterns by Andrew King (Copper Beech: Brookfield, USA, 1998). This fun math book explores patterns in nature and in numbers through games and projects using codes, algebra, and arithmetic.

What pattern do you notice in the numbers that are given? Use what you know about this pattern to fill in the missing numbers. Explain what strategies you used to find the missing numbers.

Lesson 10: Identifying, Extending, and Creating Number Patterns

Chapter Review

1. Name and describe all of the polygons you see in this design:

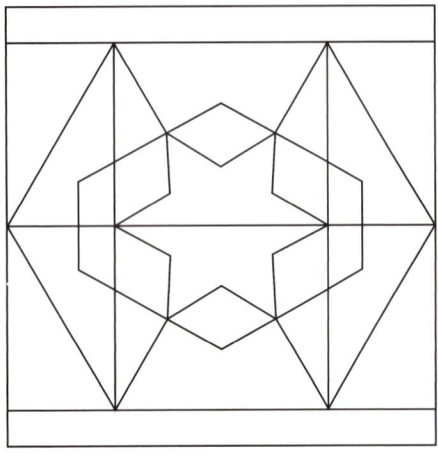

2. Draw:
    a) a right angle
    b) an obtuse angle
    c) an acute angle
    d) a pair of perpendicular lines
    e) a triangle that is both right-angle and isosceles

3. a) What type of triangle has these measurements?
      Side lengths: 8.2 m, 3.5 m, 8.2 m
   b) Explain why.
   c) Draw a picture that shows why three lengths of straw sometimes cannot be used to make a triangle.

4. a) Are these shapes congruent? Explain why or why not.
   b) Use a Mira to make a shape congruent to the shape shown below.

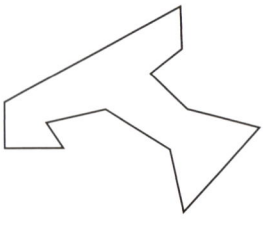 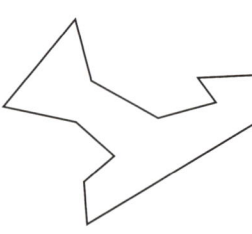

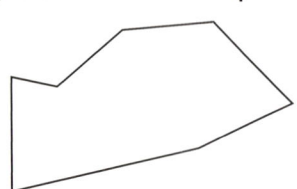

5. Complete these equivalent fractions.

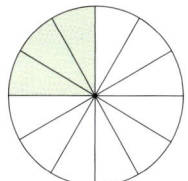 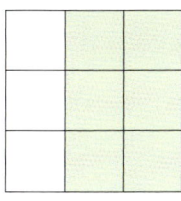 

a) $\frac{1}{4} = \frac{\blacksquare}{12}$ b) $\frac{6}{9} = \frac{2}{\blacksquare}$ c) $\frac{1}{2} = \frac{\blacksquare}{6}$

6. Name three equivalent fractions for each of these fractions:

a) $\frac{1}{2} =$ b) $\frac{3}{4} =$ c) $\frac{4}{5} =$

7. Explain how you got your answers to question 6.

8. How do you know that $\frac{2}{8} = \frac{6}{24}$? Use pictures to help support your answer.

9. Write the fraction for each shaded section.

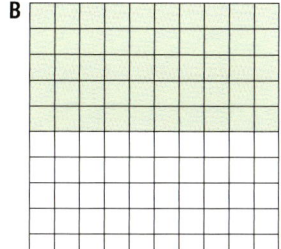

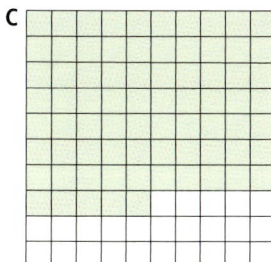 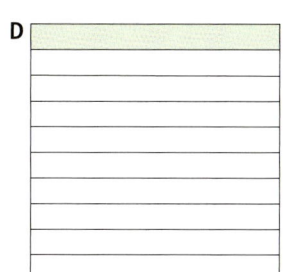

10. Write a decimal for each fraction.

a) $\frac{1}{10} = \blacksquare$ b) $\frac{6}{10} = \blacksquare$ c) $\frac{3}{10} = \blacksquare$ d) $\frac{1}{100} = \blacksquare$

e) $\frac{40}{100} = \blacksquare$ f) $\frac{28}{100} = \blacksquare$ g) $4\frac{7}{10} = \blacksquare$ h) $62\frac{49}{100} = \blacksquare$

Chapter Review

11. Write each fraction from question 10 as a decimal.

12. How do you know that 0.7 = 0.70? Use pictures to help support your answer.

13. Write a fraction for each decimal.
    a) 0.45 = ▪   b) 0.90 = ▪   c) 0.63 = ▪
    d) 0.09 = ▪   e) 0.17 = ▪   f) 0.7 = ▪
    g) 1.12 = ▪   h) 4.06 = ▪   i) 19.4 = ▪
    j) 43.43 = ▪

14. Name at least two decimals between each pair of decimals.
    a) 0.25 ▪ ▪ 0.5
    b) 0.10 ▪ ▪ 0.35
    c) 0.45 ▪ ▪ 0.60

15. Explain how you got your answers to question 14.

16. Continue each pattern.
    a) 0, 250, 500, 750, ▪, ▪, ▪
    b) 5, 2, 7, 4, 9, ▪, ▪, ▪

17. Describe the pattern rule for the patterns in 16 (a) and (b).

18. Choose three different polygons and make two different repeating patterns with them. Record your patterns and describe a pattern rule for each.

19. Make two different expanding patterns. Record your patterns and describe a pattern rule for each.

20. Where would you or your family see repeating, expanding, and/or number patterns outside of school? Record your answers.

Chapter 4: Patterns From the Past

Chapter 4 Chapter Wrap-Up

NUMBER SENSE AND NUMERATION
PATTERNING AND ALGEBRA
MEASUREMENT
GEOMETRY AND SPATIAL SENSE

In this chapter, you learned about several early civilizations. You also learned about the amazing designs people from these cultures created in their arts and crafts. Most importantly, you learned the following mathematical skills: how to identify, describe, compare, and classify two-dimensional geometric figures; how to create and compare fractions and decimals; and how to identify, expand, and create patterns, and analyze pattern rules.

Your task now is to apply what you learned about geometry, fractions, decimals, and patterns to create an attractive design for a 20 cm by 20 cm decorative tile.

Tip

Begin by making a list of important ideas and skills you learned in the lessons. Also, list tools and technology you found useful in your geometry, number, and pattern work. These lists will help you plan and create your tile. Make a rough copy of your tile on scrap paper before starting your final copy.

Traditional Kente cloth from Ghana

Decorations from National Theatre Hall, Taipei

1. Measure a 20 cm by 20 cm square on a piece of paper to represent the tile. Divide the tile into quarters.

2. Create a unique and interesting pattern in each quarter of the tile. You should have at least one repeating and one expanding pattern among your four patterns.

A pair of Mi'kmaq beaded hide moccasins

3. Include at least three polygons in your patterns. For this project, different types of triangles such as equilateral, isosceles, and scalene triangles can be counted as different polygons.

4. Include a range of geometric shapes, angles, and line segments (straight or curved).

5. Add a variety of shape sizes and colours to provide interest.

6. Imagine you are an archeologist creating a time capsule for archeologists and mathematicians of the future. In a few well-written paragraphs, explain all the pattern rules shown in your tile design. Accurately use mathematical words from this chapter. Some words you might include are polygon, acute, obtuse, repeating or expanding patterns, congruent, and sequence.

Use the checklists below to help you create your tile pattern and written explanations.

The Tile Design

- Have I been accurate in measuring and dividing my tile square into quarters?
- Have I included repeating and expanding patterns?
- Have I used a variety of polygons and other geometric shapes?
- Have I used lines and angles to add pattern variety?
- Have I used variety in sizes of shapes?
- Have I used colour effectively?
- Is my design creative and interesting?

Written Description

- Have I used mathematical vocabulary correctly?
- Are my pattern rules explained clearly? completely? accurately?
- Did I proofread my paragraphs for spelling, grammar, and clarity?

Technology

Use a paint application to complete the 20 cm² tile, using the appropriate tools. Insert text frames around the outside of the tile to explain the mathematical aspects of your patterns and shapes.

Chapter 5

Math in Ancient Cities

The people who designed and built the cities, buildings, and monuments of past civilizations needed mathematical skills to complete their work. In this chapter, you will learn the math skills necessary to create a site map that describes the shapes and sizes of structures in an ancient community.

In this chapter, you will

- estimate and measure distance and dimensions
- select the most appropriate units for different measurement situations
- estimate and calculate the perimeter and area of regular and irregular shapes
- solve problems involving measurement, perimeter, and area

Look at the map of Sacsaywaman Archeological Park below. How might you find the park's perimeter? Which unit of measurement might you choose to describe the park's size? Why?

Sacsaywaman Archeological Park, Peru

Chapter 5: Math in Ancient Cities 217

Lesson 1

Measurement—Kilometres

JOURNEY NOTE

PLAN:
You will estimate long distances in kilometres and solve problems involving distances measured in kilometres.

DESCRIPTION:
Imagine you could walk the length of a piece of history. How far would you travel? How long would it take? Construction of the Great Wall of China began around 221 B.C. and lasted over 1800 years. In this lesson you'll learn how to answer the questions above as you explore the length of the Great Wall of China.

The Great Wall of China

Get Started

How long is a kilometre? Estimate how far a kilometre stretches from your school. How many kilometres is it from your home to the school? What is the greatest distance you have ever travelled? Share your answers with the class.

1. Estimate how long it would take you to walk 100 km. What information would you need to make a reasonable estimate?

Journal

Make a list of distances in kilometres to places you regularly visit. For example, "It is 3 km to the recreation centre" or "We travel 90 km to visit Grandmother in Nanaimo." How can knowing these distances help you make other estimates in kilometres?

Vocabulary

kilo: A prefix meaning a group of 1000
kilometre: A unit for measuring distance equal to 1000 m. The metric symbol for kilometres is km.

Chapter 5: Math in Ancient Cities

2. Work with a partner. Under good conditions, the average person can walk about 4 km in an hour. How many hours do you think you could walk in a day? How long would it take you to walk 100 km? Compare your calculations with your estimate. Be prepared to discuss with the class your estimate, problem-solving strategy, and answer.

Build Your Understanding

Solve Distance Problems

According to many sources, the Great Wall of China is about 6400 km long and weaves across the northern border of China. The Great Wall varies from 5 to 15 m in height and ranges from 4 m wide at the top to 9 m wide at the bottom.

Working with a partner, find out how many days it would take you to walk the entire length of the Great Wall of China. How many hours do you think you could walk each day on such a journey? How many rest days would you need? Record your calculations.

Tip

How is this distance problem similar to the problem you solved earlier, about walking 100 km? What parts of the problem are different? How could you use what you learned from the earlier problem to solve the Great Wall of China problem?

What Did You Learn?

1. Meet with a small group of classmates. Take turns explaining how you determined the number of days your hike along the Great Wall would take. Share with the group the operations (adding, subtracting, multiplying, dividing) and strategies you used to solve the problem.

2. Make a note in your journal of any tips learned from classmates that would help you improve your own problem-solving strategies.

Practice

Math Problems to Solve

Use pictures, numbers, and words to answer the following questions:

1. A car can travel much faster than a person can walk. If a car travels at 60 km/h, how far will it travel in 5 h? in 10 h? in 20 h?

2. How long would it take for the car to travel 120 km? 180 km? 210 km?

3. How long would it take the car to travel the length of the Great Wall of China?

4. In 1980, Terry Fox wanted to run across Canada to raise funds for cancer research, so he began his Marathon of Hope. Terry ran over 5300 km, from St. John's, Newfoundland and Labrador, to Thunder Bay, Ontario. He kept up a pace of nearly 40 km per day and raised $25 million. Travelling at this rate, how long would it take him to cover 5300 km? Show your answer in pictures, numbers, and words.

Technology

Use a map, an atlas, or other resources to research the distances between some cities or towns that you know. Use desktop publishing or word processing software to record the distances in a chart. Create some math questions or word problems related to your chart. Trade with another student and answer each other's questions.

Tip

Remember that 60 km/h means the car travels 60 km in 1 h.

Terry Fox

MEASUREMENT

Lesson 2
Estimating and Measuring Dimensions of Objects

JOURNEY NOTE

PLAN:
You will learn how to select the most appropriate unit of measurement to estimate and measure the dimensions of an object. Throughout the lesson you will practise and develop your estimating and measuring skills.

DESCRIPTION:
Archeologists often dig to find artifacts, or items used by people of early civilizations. They might uncover an entire object, or parts of an object. Archeologists measure the items or pieces they find and keep accurate records of sizes and exact locations where the items are found.

An archeological dig

Get Started

When a kilometre is too large a unit of measurement, metres, centimetres, or millimetres can be used.

Journal

Make a glossary of measurement prefixes. Most dictionaries give definitions of the prefixes. Write glossary entries in your own words.

Vocabulary

centi: A prefix meaning one hundredth ($\frac{1}{100}$)
centimetre: One hundredth of one metre or 0.01 m. The metric symbol for centimetre is cm.
dimension: A measurement such as the height, width, length, or depth of an object
metre: A unit of measurement used to measure length. The metric symbol for metre is m.
milli: A prefix meaning one thousandth ($\frac{1}{1000}$)
millimetre: One thousandth of one metre or 0.001 m. The metric symbol for millimetre is mm.

Lesson 2: Estimating and Measuring Dimensions of Objects

The dimensions of these sites and artifacts from early civilizations are best given in metres:

The Serpent Mound, in Ohio, is a 390-m twisting mound that resembles the shape of a snake. It was built by the Adena people more than 2000 years ago. The Serpent Mound averages about 1.5 m in height and about 6 m in width. Snakes played an important role in the beliefs of the Adena people. Mounds made to look like snakes can also be found in Ontario.

In about 300 B.C., King Philip of Macedon had his soldiers carry spears that were 5 m long.

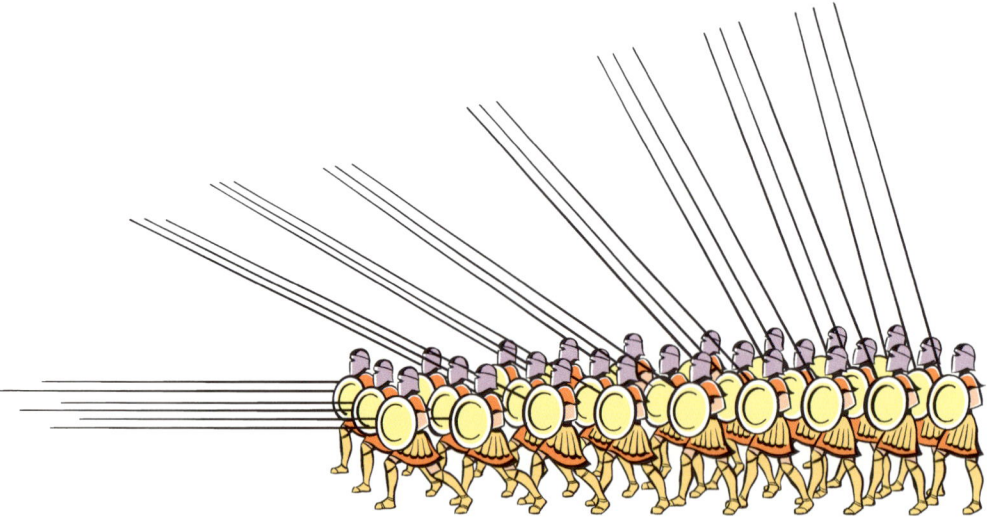

1. In a small group, discuss why metres are the best unit of measurement for describing the objects shown above.

Small items require smaller units of measurement. In your notebook, answer the following questions.

2. The gold Aztec necklace is best measured in centimetres. The width of an ancient gold earring is best measured in millimetres. Why?

3. Find something in the classroom that you estimate is 1 m long, something that is 1 cm long, and something that is 1 mm long. Check your estimates by using a ruler to measure the objects you selected. Were metres, centimetres, or millimetres the best measurement unit in each case? Why? If not, what would be a better unit of measurement for the object? Why?

4. Identify other objects in the classroom that would best be measured in metres, centimetres, and millimetres.

Greek gold necklace

Gold earrings from an early Mediterranean civilization

Tip

In most classrooms, a metre is about the height of a doorknob from the floor. A finger's width is usually about a centimetre. A small paper clip is about a millimetre thick.

Build Your Understanding

Estimate and Measure Dimensions

You Will Need
- ruler

1. Copy this chart:

Item:		Unit(s) of Measurement:	
Dimension	Estimate	Actual Measurement	
Length			
Width			
Height of Object			

Lesson 2: Estimating and Measuring Dimensions of Objects

2. For each item you selected in question 4 of Get Started, record its name and the unit of measurement you selected. Then, estimate each dimension of the object and record your estimate on the chart. You may need to use different measurement units for each dimension, depending on your chosen objects.

3. Use a ruler to accurately measure the object's dimensions and record your results. Round to the nearest unit.

4. Choose three other objects: one that is best measured in metres, one in centimetres, and another in millimetres. Repeat the above activity, estimating the dimensions of each object, then measuring it. Make sure to record your estimates and measurements.

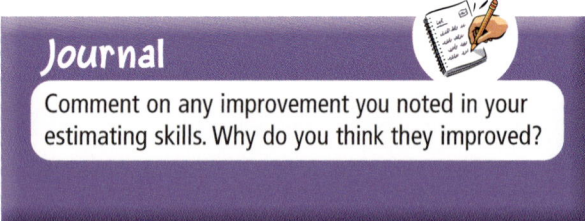

Journal

Comment on any improvement you noted in your estimating skills. Why do you think they improved?

What Did You Learn?

1. Share with the entire class items you found that are best measured in metres, centimetres, or millimetres. Organize group findings into a class chart.

2. Describe any challenges you faced when measuring your items.

3. Do you find it easier to estimate dimensions in metres, centimetres, or millimetres? In a few sentences, explain why. Compare and discuss your reasons with a classmate.

4. How do you decide what the best unit is for measuring the dimensions of an object? Give examples to support your answer.

Practice

What would be the best unit for measuring each of the following objects' dimensions? Why?

1. the length of a tennis court
2. the distance across Canada
3. the length of a pen
4. the width of a pen
5. the thickness of a $1.00 coin
6. Compare and discuss your answers for questions 1 to 5 with a classmate. Give reasons for each measurement decision.
7. Trace the outline of your shoe on a large piece of scrap paper. Estimate the length of your shoe and the width at the widest part. What units did you use? Why? Record your estimates. Use a ruler to accurately measure each dimension.
8. What would be the best unit for measuring the dimensions of a chalkboard? Why? Estimate then measure the length and width of a chalkboard in your classroom. Compare the actual results with your estimate.
9. Use an on-line fact book to find the heights of the 10 largest buildings in the world. Using a computer application, record this data on a spreadsheet. Compare these heights to the CN Tower. Display your findings on a bar graph.

Technology

Use a draw application to create a collage of items grouped by the metric unit used to measure them. The display could include stamps or drawings of items, or words and phrases. Include both ancient artifacts and present-day items.

Lesson 3
Relationships Between Metres and Centimetres

JOURNEY NOTE

PLAN:
In a group activity, you will use a metre stick to measure the height of all group members, then write the heights in centimetres; metres and centimetres; and metres with decimals.

DESCRIPTION:
Many hundreds of years before European explorers arrived in North America, native peoples had thriving civilizations. In West Coast native communities, many houses had totem poles called "house poles," which might have been over 1 m wide at the base and over 15 m in height.

A West Coast native community

Get Started

You Will Need
- metre stick

You will explore the relationship between two metric units—metres and centimetres.

The basic unit for measuring length or height is the metre. The word "metre" comes from the ancient Greek word "metron," which means a measure. All of the other distance units in the metric system are related to the metre. For example, a centimetre is $\frac{1}{100}$ of a metre and a millimetre is $\frac{1}{1000}$ of a metre. All of these standard measurement units are part of the metric system, which is about 200 years old.

Journal

Make a web of all the words you know containing the word "metre"; for example, "kilometre" and "centimetre." Circle the prefix of each web word. Add new words as you come across them in the chapter and in any measuring you do outside the classroom.

1. Work with a partner. Examine the markings on a metre stick. How long is it? How is it divided? What does each type of marking on the metre stick mean? How do you know?

Chapter 5: Math in Ancient Cities

2. How many centimetres are in a metre? How did you reach that conclusion?

   A centimetre can be written as a fraction and as a decimal of a metre:

   1 cm = $\frac{1}{100}$ m or 0.01 m

   Metres and centimetres can also be used together.

3. Find something in your classroom with a dimension that is best given in metres *and* centimetres. Write your measurement.

4. Next, write the dimension from question 3 in the following ways:

   a) in metres with centimetres given as a fraction of a metre

   b) in metres with centimetres given as a decimal of a metre

   c) completely in centimetres

   Which way is best? Why?

5. What is the length of the floor in your classroom? Write your answer using the three ways listed in question 4. Explain which way is best.

6. Visit the Canadian Museum of Civilization Web site at www.civilization.ca to learn more about the totem poles erected by the West Coast native peoples. Why were they carved? In what ways are they similar, and in what ways are they different from one another?

The world's largest free-standing totem pole is in Victoria, British Columbia.

Tip

You might create conversion charts for changing centimetres into fractions or decimals of a metre. For example, your first chart entry could be
1 cm = $\frac{1}{100}$ m or 0.01 m.

Lesson 3: Relationships Between Metres and Centimetres

Build Your Understanding

Measure and Compare Heights in Metres and Centimetres

You Will Need
- metre stick or measuring tape
- calculator (optional)

In 1991, two hikers in the Alps discovered the frozen body of an ancient man believed to be a shepherd or a hunter. The man was named Iceman or Ötzi (after the mountains where he was found). Iceman is believed to be 5300 years old, and he lived during the late Stone Age. He was 160 cm tall. How does Iceman's height compare with your height and the height of your classmates?

1. Work in small groups. Make a copy of this chart in your notebook. The "Name" column should include the names of all students in your group.

Name	Height in m and cm	Height in cm	Height in m With Decimals
Iceman	1 m 60 cm	160 cm	1.60 m

2. Measure and record the heights of all group members in metres and centimetres.

3. Write all heights in centimetres only in the correct chart column.

4. Write all heights in metres only, expressing parts of a metre as a decimal.

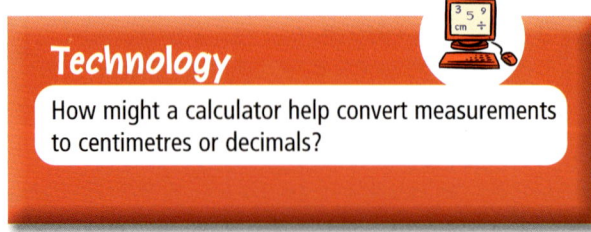

Technology
How might a calculator help convert measurements to centimetres or decimals?

Chapter 5: Math in Ancient Cities

What Did You Learn?

1. Meet as a class and share your results. Use these results to create a class comparison chart.

2. How does your height compare with that of Iceman? How do the heights of your group's members compare? How do the heights of all students in your class compare?

3. Explain how you converted your height from metres and centimetres to only centimetres.

4. Explain how you converted your height from metres and centimetres to metres only, expressing parts of a metre as a decimal.

5. It is estimated that Iceman was 50 years old when he died. Are some of the students in your class already taller than Iceman? If so, why might that be?

6. What other conclusions can you draw from the class height data?

Practice

1. Convert the following measurements into metres. Write parts of a metre as a decimal.
   a) 456 cm
   b) 110 cm
   c) 906 cm
   d) 3 m 46 cm
   e) 12 m 50 cm

2. Order the above measurements from smallest to largest. Explain how you know your order is correct.

3. Research Yukon Man, Ice Maiden from Siberia, or another discovery of a human body that has helped modern people understand very early peoples. Include in your research notes any important measurements; for example, the height of the person or how far he or she was found from a major modern city. Share your findings with the class in a short oral or written report. Explain why you used the units of measurement you did.

Lesson 4

More Relationships Between Units of Measurement

JOURNEY NOTE

PLAN:
In this lesson, you will practise estimating and measuring skills. You will strengthen your understanding of metric unit prefixes, and you will explore the relationship among kilometres, metres, decimetres, centimetres, and millimetres.

DESCRIPTION:
The ancient Egyptians used a unit of measurement called the Royal Cubit, which was equal to 52.4 cm. The Royal Cubit was the standard length that was used to measure buildings, including the dimensions of the Great Pyramid.

The Great Pyramid at Giza

Get Started

In Lesson 3, you learned about the relationship between a metre and a centimetre. You will recall that 1 cm = $\frac{1}{100}$ m. Adding prefixes to the word "metre" can create smaller or larger units of measurement.

The following are important measurement prefixes:

kilo: 1000   **deci:** $\frac{1}{10}$

centi: $\frac{1}{100}$ **milli:** $\frac{1}{1000}$

The prefixes above can be added to "metre" to create these words for units of measurement:

kilometre   decimetre
centimetre   millimetre

The strip below is ten millimetres long. The distance between each mark is one millimetre. The metric symbol for millimetre is mm.

The strip below is one centimetre long.
The metric symbol for centimetre is cm.

The strip below is one decimetre long.
The metric symbol for decimetre is dm.

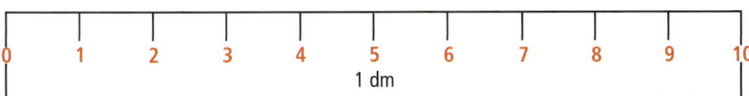

You Will Need
- scrap paper for tracing
- scissors
- one metre stick per group

1. Trace and cut out the decimetre, centimetre, and ten-millimetre strips shown above. Check the length of each strip with your ruler. Make sure the millimetre marks are clear on the 10-mm strip. Label each strip.

2. Work in small groups so you can share materials. Compare the strips by placing them side by side to find out, for example, how many centimetres are in a decimetre, how many millimetres in a centimetre, and so on. Experiment with as many comparison possibilities as you can. Record your answers.

3. Find five small objects in the classroom that you estimate are no longer than a metre. The objects must range widely in length. Measure each object in decimetres, centimetres, and then millimetres.

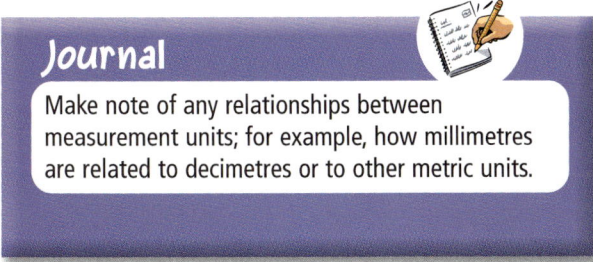

Journal

Make note of any relationships between measurement units; for example, how millimetres are related to decimetres or to other metric units.

Lesson 4: More Relationships Between Units of Measurement

You can use your cut-out patterns and a metre stick to explore the relationships between metric units; for example:

1 dm = ■ m   Answer: 1 dm = $\frac{1}{10}$ m or 0.1 m

4. Use the cut-out patterns and a metre stick to answer the following questions about relationships between metric units:

   a) What fraction and decimal of a metre is a decimetre?
      1 dm = ■ or ■ m

   b) How many centimetres in one decimetre?
      1 dm = ■ cm

   c) What fraction and decimal of a decimetre is a centimetre?
      1 cm = ■ or ■ dm

   d) How many millimetres in one decimetre?
      1 dm = ■ mm

   e) What fraction and decimal of a decimetre is a millimetre?
      1 mm = ■ or ■ dm

   f) How many centimetres are there in one metre?
      1 m = ■ cm

   g) What fraction and decimal of a metre is a centimetre?
      1 cm = ■ or ■ m

   h) How many millimetres in one centimetre?
      1 cm = ■ mm

   i) What fraction and decimal of a centimetre is a millimetre?
      1 mm = ■ or ■ cm

   j) How many millimetres are there in one metre?
      1 m = ■ mm

   k) What fraction and decimal of a metre is a millimetre?
      1 mm = ■ m or ■

Tip

Use a calculator to help you answer the above questions.

Vocabulary

centimetre: $\frac{1}{100}$ of a metre or 0.01 m
decimetre: $\frac{1}{10}$ of a metre or 0.1 m
millimetre: $\frac{1}{1000}$ of a metre or 0.001 m

Chapter 5: Math in Ancient Cities

Build Your Understanding

Estimate and Measure Using Millimetres, Centimetres, and Decimetres

1. Work with a partner. Estimate in millimetres the length, width, and height of your mathematics textbook and your mathematics notebook.

2. Record your estimates in a chart like this.

Dimension		Estimate in mm	Actual Measurement		
			mm	cm	dm
Text	Length				
	Width				
	Height				
Notebook	Length				
	Width				
	Height				

3. Next, measure the dimensions of both books in millimetres, centimetres, and decimetres. Record your measurements in the appropriate spaces on the chart.

What Did You Learn?

1. Describe in pictures, numbers, and words the relationships between the following:

    a) a millimetre and a centimetre

    b) a millimetre and a metre

    c) a decimetre and a centimetre

    Share, compare, and discuss your explanations of the relationships with a classmate.

2. Which measurement unit do you think is best for measuring the dimensions of your mathematics books? Why? Compare and discuss your conclusion with a small group of classmates.

Lesson 4: More Relationships Between Units of Measurement

Practice

Name the best unit for measuring the following. Explain your choices.

1. the length of a car
2. the thickness of a library card
3. the height of the Great Pyramid in Egypt
4. the length of a pen
5. the distance from Toronto to Regina

Use what you know about the relationships among metric units to complete the following statements:

6. 2 dm = ■ cm or ■ mm
7. 3 m = ■ dm or ■ cm or ■ mm
8. 6 cm = ■ mm

Convert the first measurement to a fraction and then a decimal.

9. 3 mm = ■ cm or ■ cm
10. 4 cm = ■ dm or ■ dm

11. You learned in the introduction that the ancient Egyptians used a unit of measurement called a Royal Cubit, which is equal to 52.4 cm in our metric measurement system.

    The Egyptians divided each Royal Cubit into 7 equal "palms." Each palm was divided into 4 "digits." Work in a group. Use a long piece of paper to make a Royal Cubit ruler. Then use your Royal Cubit ruler to measure the dimensions of your classroom.

Show What You Know
Review: Lessons 1 to 4, Measurement

1. Canada has the world's longest recreational trail. It is the Trans Canada Trail, which is about 16 000 km long. At your own rate of walking, how long would it take you to walk the entire trail? What factors might affect how long it took you to walk this distance? Use pictures, numbers, and words.

2. Estimate then measure the length of a door in your classroom or school. Record the actual dimension in as many ways as you can.

3. Describe the relationship between a decimetre and a metre. Use pictures, numbers, and words.

Lesson 5

Calculating Perimeter

JOURNEY NOTE

PLAN:
You will learn or review the rules for calculating the perimeters of rectangular and square figures, and other polygons. Then you will solve problems that involve calculating the perimeter of two-dimensional shapes.

DESCRIPTION:
In early civilizations, many cities were surrounded by walls to protect the inhabitants from enemies. People who designed and built the walls needed to know the distance around the entire city. This distance is called the perimeter.

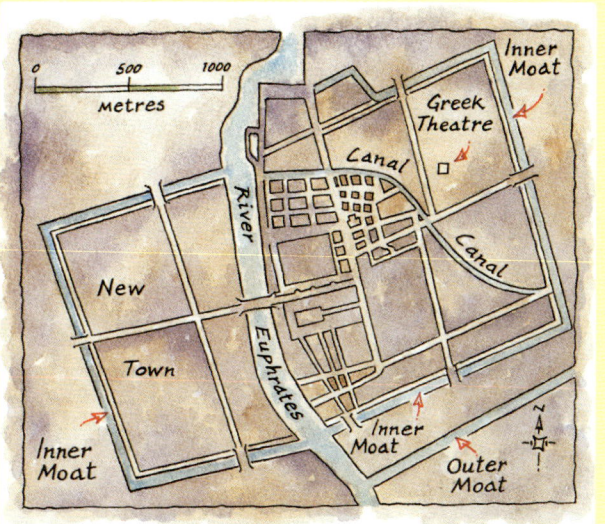

A plan of the ancient city of Babylon

Get Started

Perimeter is the distance around a figure such as an object, building, part of a building, or a region. What strategies could you use to determine perimeter? Brainstorm different approaches that may be used, and share them with the class.

If you want to find the perimeter of a four-sided figure, such as the closet door to the right, add up all the dimensions.

A student calculated the perimeter by adding the two lengths and widths of the rectangular closet. 1 + 2 + 1 + 2 = 6. The perimeter of the closet is 6 m. Is there another way you can calculate the perimeter of the closet?

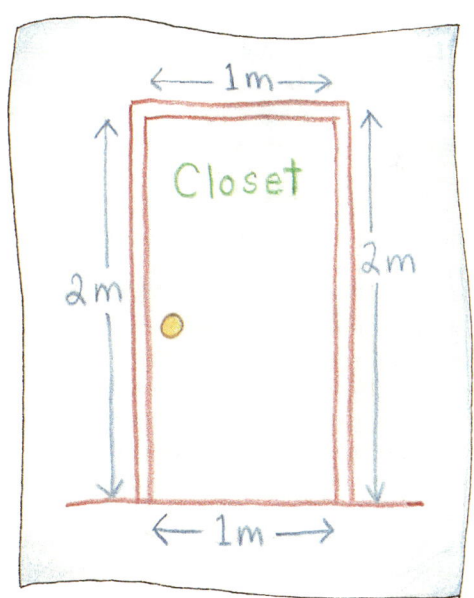

Vocabulary

perimeter: The distance around a figure

Build Your Understanding

Find the Perimeter of a Schoolroom, Building, or Region

You Will Need
- measuring tool

Work in small groups to find the perimeter of a room, building, or region at your school.

1. Select the room, building, or region. What metric units will you use to measure the figure? Why? What measuring tools will be most useful? How will you make your measurements as accurate as possible? Write out a simple, clear project plan.

2. Look closely at the room, building, or region and draw a rough sketch of its shape. Estimate the dimensions of the shape. Record your estimates on the sketch.

3. Measure each dimension of the room, building, or region you selected. Record the dimension measurements on the sketch in a different colour from the one you used to record estimates.

4. Calculate the perimeter of the room, building, or region you selected.

5. Use your rough sketch to make a more accurate drawing of the room, building, or region. On your final drawing, write a title and record the lengths of each side of the figure and the perimeter.

Tip
You might show a relationship between the unit of measurement in your drawing and the unit of measurement in the real-world object. For example, 1 cm on your drawing could equal 1 m on the real-world object. This is called the "scale." It is written 1 cm = 1 m.

Journal
Outline the steps you followed to find the perimeter of a part of the school. Could you use this strategy or rule again to find the perimeter of other similar-shaped figures? If so, explain how.

What Did You Learn?

1. Meet as a class and display your perimeter drawings.

2. Make a list of all the school rooms, buildings, and regions that class members measured. You can organize the drawings to create a school plan or map.

3. Identify parts of the school that have similar perimeters. How do the dimensions of these similar parts compare?

4. Name other parts of the school that class members did not measure that you estimate have similar perimeters. Check your estimates.

5. Did you use the best metric unit to estimate and measure the part of the school you selected? Why? If not, name a better unit of measurement. Give reasons for your conclusion.

6. Summarize the rules for calculating the perimeters of rectangles and squares.

Journal

Make some notes on what you learned about estimating and finding perimeters as a result of sharing and discussing ideas with classmates.

Booklink

Eyewitness Books: Building by Philip Wilkinson (Stoddart: Toronto, ON, 1995). This book presents details about all kinds of structures, from African mud huts to Japanese skyscrapers to Turkish mosques.

Practice

Extension

1. Draw a rectangle that is 6 cm long and 5 cm wide. What is its perimeter?

2. Draw a square with a side length of 4 cm. What is its perimeter?

3. Calculate the perimeter of the figures shown below. Show your work.

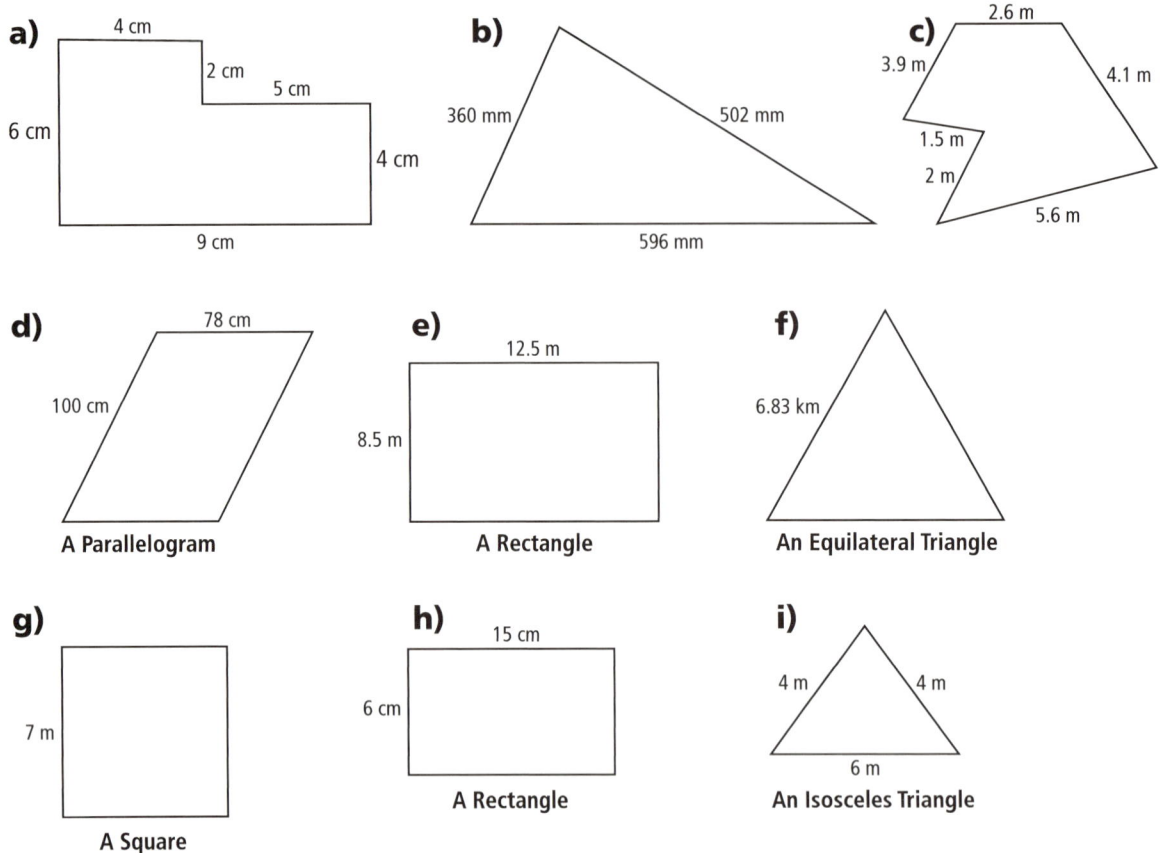

a) Irregular shape: 4 cm, 2 cm, 5 cm, 6 cm, 4 cm, 9 cm

b) Triangle: 360 mm, 502 mm, 596 mm

c) Irregular polygon: 2.6 m, 3.9 m, 4.1 m, 1.5 m, 2 m, 5.6 m

d) A Parallelogram: 78 cm, 100 cm

e) A Rectangle: 12.5 m, 8.5 m

f) An Equilateral Triangle: 6.83 km

g) A Square: 7 m

h) A Rectangle: 15 cm, 6 cm

i) An Isosceles Triangle: 4 m, 4 m, 6 m

4. The ancient Sumerians built incredible temples called "ziggurats."

The base of the ziggurat was a rectangle that measured 65 m by 45 m. Find the perimeter of the ziggurat's base. Show your work.

Architects and architectural technicians need to know the dimensions and perimeters of rooms and buildings quickly. Computers can help them obtain this information by doing necessary calculations very fast.

The ziggurat at Ur, as it looked when it was newly built

Chapter 5: Math in Ancient Cities

MEASUREMENT
GEOMETRY AND SPATIAL SENSE

Lesson 6
Perimeter—Irregular Shapes

JOURNEY NOTE

PLAN:
You will calculate the perimeter of irregular shapes, and you will solve problems that involve calculating the perimeter of irregular, two-dimensional shapes.

DESCRIPTION:
The pharaohs, or rulers of ancient Egypt, wrote their names in hieroglyphics, a form of picture writing. The name, enclosed in an oval, has come to be called a "cartouche."

An ancient Egyptian cartouche

Get Started

In Lesson 5, you learned how to calculate the perimeter of regular shapes such as squares. In this lesson, you will learn how to find the perimeter of an irregular shape. An irregular shape has unequal sides and angle measures.

1. With a partner, estimate the perimeter of the irregular shape to the right. Then measure and calculate the perimeter.

2. Meet with another pair to compare your results and the strategies you used to calculate the perimeter.

Vocabulary

irregular shape: A shape with unequal sides and angle measures

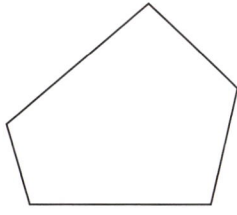

Tip

Look at the strategy you used to calculate the perimeter of a regular shape, such as a square. How is the irregular shape different from the regular shape? How will this affect the way you calculate the perimeter?

Lesson 6: Perimeter—Irregular Shapes

How would you find the perimeter of the oval part of the cartouche to the right?

You Will Need
- ruler
- string
- measuring tape
- any other measuring tools you think will be helpful

3. Work with a partner. Estimate the perimeter of the cartouche. Record your estimate.

4. Outline the strategies and measurement tools you will use to measure the perimeter.

5. What is the best metric unit to use when measuring this cartouche? Why?

6. Measure and calculate the perimeter of the cartouche.

Tip

Think about how you will measure the straight part of the shape. How will you measure the curved part? What materials or tools might help you obtain an accurate measure of the curves?

Build Your Understanding

Measure the Perimeter of an Ancient City

The city of Ur was built more than 4800 years ago in Mesopotamia.

1. With a partner, estimate the perimeter of the map of the city of Ur, to the right. Record your estimate.

2. Measure and calculate the perimeter of the city. How close was your estimate?

3. If the scale of the diagram of Ur is 1 cm = 100 m, what is the perimeter of the actual city of Ur? Explain how you got your answer.

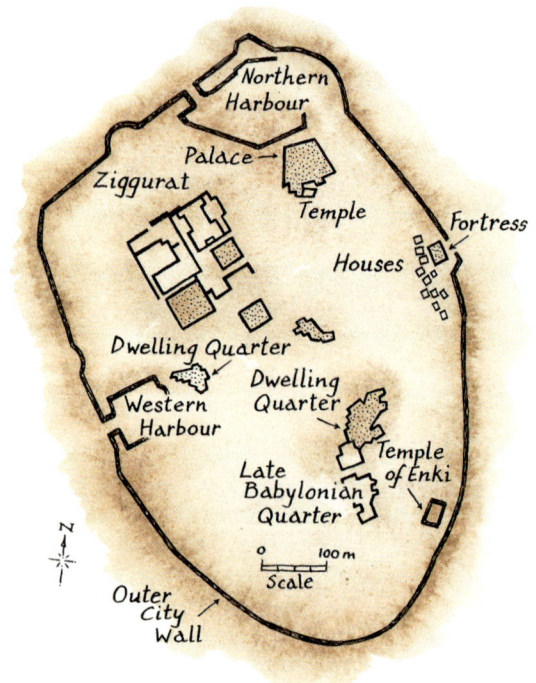

What Did You Learn?

1. Meet with another group. Compare your perimeters of the cartouche and the ancient city. Which method for finding the perimeter do you think was most accurate? Why?

2. Which metric unit was best for measuring the perimeter of the cartouche? Why?

3. Which metric unit was best for measuring the perimeter of the ancient city of Ur? Why?

Journal
Explain the method or strategy you used to find the perimeter of irregular shapes.

Practice

Extension

1. The walls surrounding the city of Ur were 37 m thick. Estimate then measure the hallway of your school. How many "hallways" thick were the walls that protected Ur? Show your work.

2. Use a search engine to find a Web site that displays the English alphabet and corresponding hieroglyphic symbols. Design and create a cartouche of your name using a paint application or by hand.

   Calculate the perimeter of your cartouche. Exchange the cartouche with a classmate and challenge him or her to estimate and then calculate the perimeter of the shape.

Show What You Know
Review: Lessons 5 and 6, Perimeter

How would you explain perimeter to someone who doesn't know what it is? Give explanations for regular and irregular shapes. Use pictures, numbers, and words.

MEASUREMENT
GEOMETRY AND SPATIAL SENSE

Lesson 7

Finding the Area of Irregular Shapes

JOURNEY NOTE

PLAN:
You will estimate the area of irregular shapes using grid paper. You will find the area of irregular shapes using 1-cm grid paper and a 1 m by 1 m square.

DESCRIPTION:
The cities of early civilizations, such as ancient Rome, were built in various shapes and sizes.

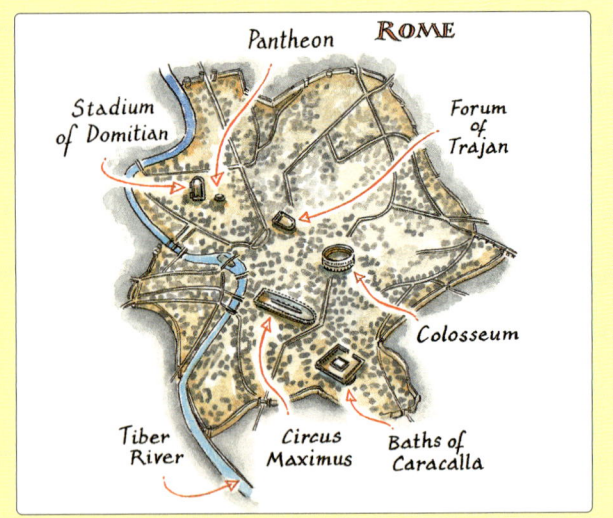

A map of ancient Rome

Get Started

The area of a figure is a measure of the surface the figure covers. Area is measured in square units, such as square centimetres (cm^2) or square metres (m^2).

A square centimetre is the amount of surface covered by a square that is 1 cm long on each of its four sides.

The shape to the right does not have regular sides. Estimate the area of the shape and record your estimate.

How can you find the area of this shape? Use the grid legend above the shape to find the area. With a partner, compare the strategies you used to calculate the area.

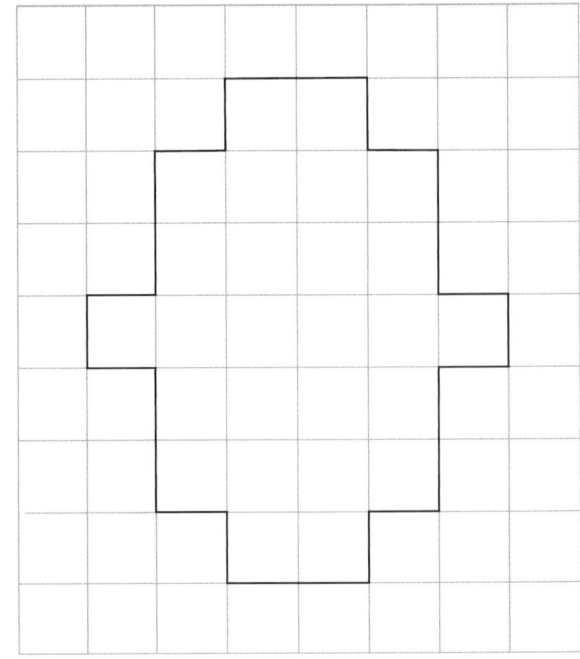

242 Chapter 5: Math in Ancient Cities

Work with a classmate. Estimate the area of the shape below. What information did you use to make your estimate? How can you find the area of the irregular shape below?

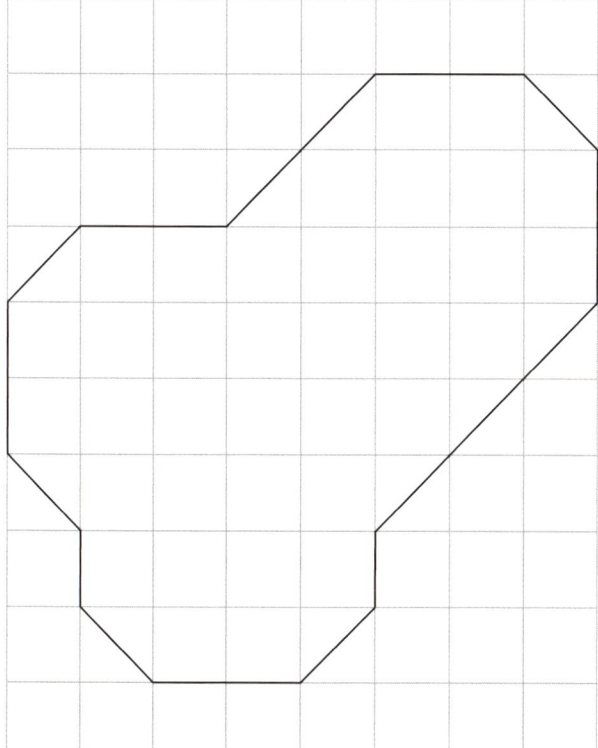

Tip
The grid on which the shape appears is made up of squares that are 1 cm x 1 cm, or 1 cm². When counting, develop a strategy for keeping track of incomplete squares.

Journal
Explain the strategy you used to find the area of the irregular shapes on grid paper.

Vocabulary
area: The amount of surface inside a two-dimensional shape. Area is measured in square units: mm², cm², dm², m², km². Square units are units that measure equally on all four sides.
grid paper: Paper marked off in equal-size squares. For example, 1 cm x 1 cm, or 1 cm².

Use the legend on the previous page to find the area of the irregular shape.

Build Your Understanding

Measure in Square Metres

How can you measure the area of large irregular shapes? Sometimes square centimetres (cm²) are not the best measurement unit. You may need to use square metres (m²).

You Will Need
- scissors
- cardboard
- metre stick

1. Work with a partner to make a one-metre-square piece of cardboard.

Lesson 7: Finding the Area of Irregular Shapes 243

2. Find an irregular shape within your classroom or school. Estimate the area of the shape you selected in square metres (m²), and record your estimate.

3. Make a simple plan that explains how you will use your metre-square piece of cardboard to measure the area of the irregular shape. Record your plan.

4. Use your metre-square piece of cardboard to measure the area of the irregular shape in square metres (m²).

5. Find another irregular shape and repeat the process of estimating and measuring. Make sure that you record your estimates and measurements.

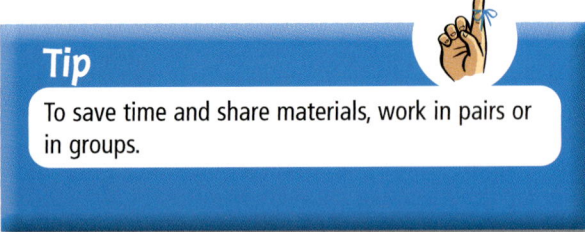

Tip

To save time and share materials, work in pairs or in groups.

What Did You Learn?

1. Have a spokesperson from your pair or small group identify the irregular shape you measured. Challenge classmates to predict the area of that shape in square metres (m²). The spokesperson will announce the actual measurement.

2. Explain the method or strategy you used to find the area of the irregular shape.

3. Were square metres the best metric unit to use when measuring the area of your shape? Explain why or why not.

4. How could you make this method more accurate?

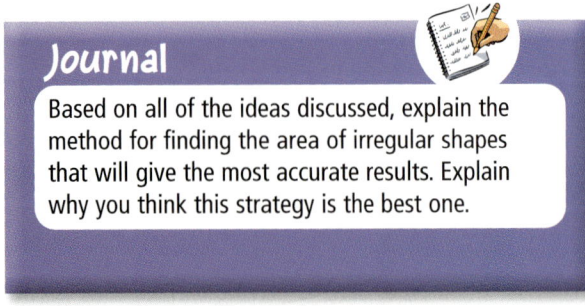

Journal

Based on all of the ideas discussed, explain the method for finding the area of irregular shapes that will give the most accurate results. Explain why you think this strategy is the best one.

Practice

1. Give the area of each shape below.

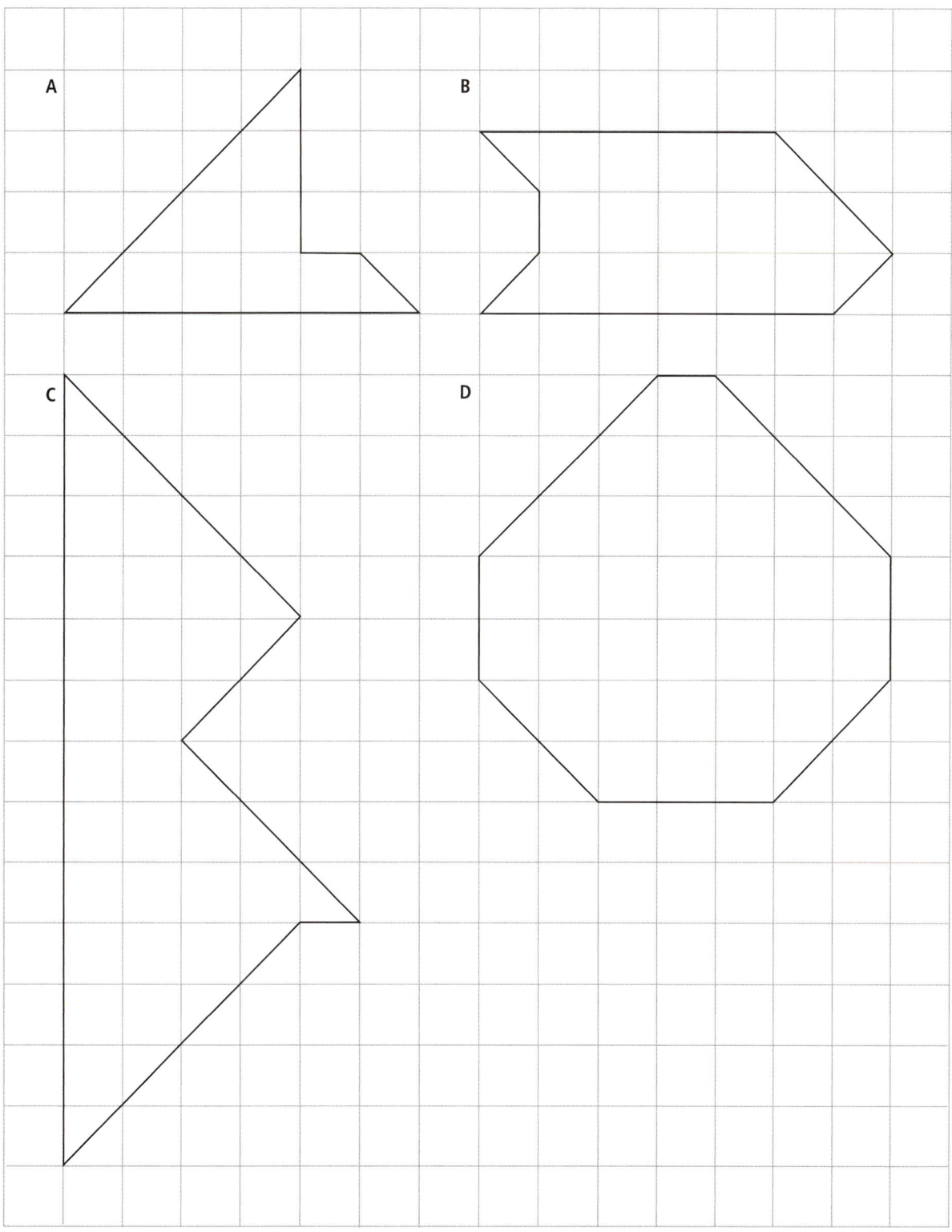

Lesson 7: Finding the Area of Irregular Shapes

2. Explain how you calculated the area for the shapes in question 1.

3. On grid paper, using straight lines, draw five irregular shapes. Find the area of each shape in square centimetres. Exchange your shape drawings with a classmate and challenge him or her to find the area of each shape.

4. How would you find the area of each shape in question 3 if the grid paper on which you drew the irregular shapes had squares measuring 0.5 cm by 0.5 cm?

5. Use a paint application with a ruled grid background. Draw and fill irregular shapes that are 26 cm^2, 33 cm^2, and 52 cm^2.

Math Problems to Solve

6. Using 1-cm grid paper, make five irregular shapes that are 15 cm^2.

7. Karen's backyard can hold a swimming pool that is 30 m^2. Use grid paper to show examples of what the pool might look like. Pretend that one square on your grid represents 1 m^2.

8. Estimate then find the area of one of your hands. Explain the method or strategy you used to find the area of your hand. Do you think your answer is accurate? Explain why or why not.

Lesson 8
Measurement Relationships and Formulas

MEASUREMENT
GEOMETRY AND SPATIAL SENSE

JOURNEY NOTE

PLAN:
You will draw squares and rectangles on grid paper, then develop rules for calculating the perimeter and area of these shapes. Next you will learn and apply perimeter and area formulas.

DESCRIPTION:
Imagine you are an archeologist studying past civilizations and you must find the perimeter and area of an ancient city. One site is a square and the other is a rectangle. You know only the lengths and widths of the city sites. How would you find the perimeter and area of these large surfaces without using huge grid paper or massive cardboard squares?

Get Started

Counting sides of squares or whole squares on a grid can take a long time when you want to find the perimeter or area of large squares or rectangles. Is there a better way to find the perimeter and area of these shapes? There is if you take time to explore the relationships between the lengths and widths of the shapes, and their perimeters and areas.

Build Your Understanding

Explore Dimensions, Perimeters, and Areas of Squares and Rectangles

You Will Need
- 1-cm grid paper

1. Work in a small group. Each person in your group draws three different squares or rectangles on a piece of grid paper. Make sure you have a good sample of both squares and rectangles. Label the shapes A, B, C, and so on.

2. Make a chart like the one below in your notebook. You will need one row for each shape created in question 1. Label each shape as a square or rectangle.

Shape	Length	Width	Perimeter	Area

3. Use the grid paper to help you find the length, width, perimeter, and area of each shape your group created. Record your results on the chart.

4. As a group, study your completed chart. What relationship do you notice between the side dimensions of squares and their perimeters? Write a statement to summarize your group's conclusions.

5. What relationship do you notice between the side dimensions of squares and their areas? Write a statement to summarize your group's conclusions.

6. What relationship do you notice between the lengths and widths of rectangles and their perimeters and areas? Write a statement to summarize your group's conclusions.

Journal

Make note of any patterns or relationships your group noticed on the chart.

Tip

Refer to your group chart and the shapes on the grid paper when making points for the discussion in the next part of this lesson.

Chapter 5: Math in Ancient Cities

What Did You Learn?

1. Meet as a class to share and discuss your drawings, chart results, and summary statements.

2. Use pictures and words to describe the relationship between the length, width, and perimeter of squares and then rectangles.

3. Use pictures and words to describe the relationship between the length, width, and area of squares and rectangles.

4. A formula is a sentence that expresses a general rule. Read the formulas for finding the area and perimeter of squares and rectangles on the right.

   Explain how the formulas compare with your relationship statements in questions 2 and 3.

Vocabulary

formula: A set of symbols, such as letters and numbers, used to give a general rule

Formulas
Squares and Rectangles

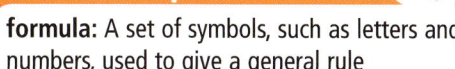

Area = length x width
$A = l \times w$

Perimeter = 2 length x 2 width
$P = 2l \times 2w$

The same formula is used for both squares and rectangles.

Practice

1. For each shape to the right, estimate its perimeter and area. Then, use the correct formula to calculate the perimeter and area.

2. Using your school library or the Internet, find the dimensions of an important city or building from an early civilization. The community or structure must be a square or rectangle. Share with a classmate the name of the city or building, the civilization that built it, and its perimeter and area.

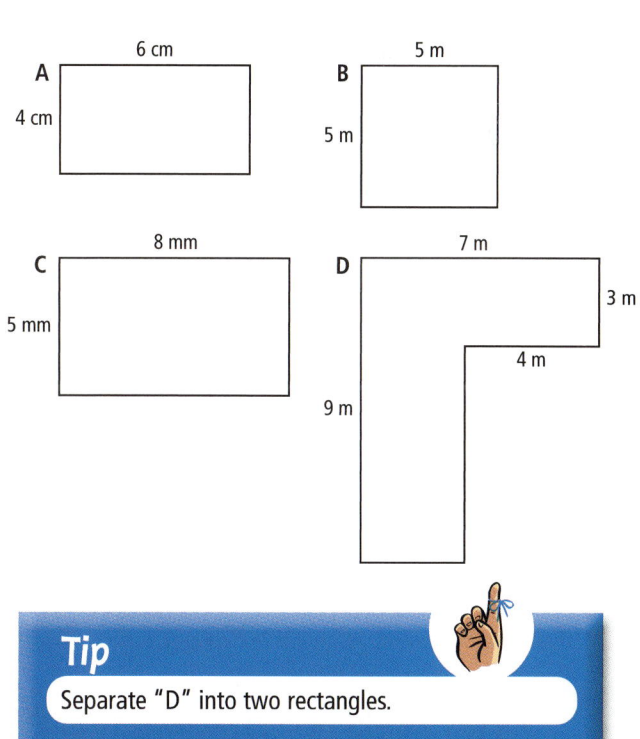

Tip
Separate "D" into two rectangles.

Lesson 8: Measurement Relationships and Formulas

Lesson 9
Changing the Dimensions of a Rectangle

JOURNEY NOTE

PLAN:
You will draw rectangles of increasing size, then estimate and calculate the perimeters and areas.

DESCRIPTION:
Tutankhamen was a famous Egyptian pharaoh. Imagine you are the designer or builder of a monument for the pharaoh, or ruler, of ancient Egypt. The pharaoh needs the length or width of one of the rectangular rooms increased by one unit of measurement. What effect will this have on the perimeter and area of the room?

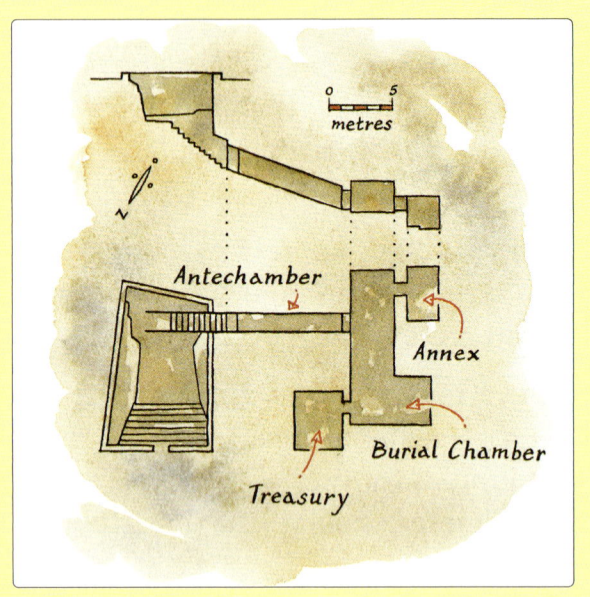

Get Started

You Will Need
- 1-cm grid paper

1. In your notebook, make a chart like the one on the right.

2. Draw a small rectangle on grid paper. Record the rectangle's dimensions on the chart.

3. Estimate the perimeter and area of your rectangle. Then, use the formulas in Lesson 8, page 249, to calculate the perimeter and area of your rectangle. Record your calculations on the chart. Check your calculations by counting square lengths or whole squares on the grid paper.

Dimensions of Rectangle	Perimeter $2l + 2w$	Area $l \times w$

Tip
Make sure that the rectangle you draw includes only complete squares on the grid paper.

Technology
Use a calculator to check your answers.

4. Draw a second rectangle, but increase one dimension, either the length or the width, by 1 cm. Record this rectangle's dimensions on the next line of the chart. Estimate the perimeter and area of the new rectangle. Then, calculate the perimeter and area of the rectangle, and record your calculations on the chart. Make note of any interesting patterns you observe in your chart results.

Build Your Understanding

Make Line Graphs of Chart Results

You Will Need
• line graph reproducible page

You are going to make line graphs of your chart results. One line graph will show what happened to the perimeter of the rectangle when you increased a dimension by 1 cm. The other line graph will show what happened to the area.

1. Work with a partner and start with the perimeter line graph. On the "Dimension" axis, record the dimensions of each of your rectangles. Record them in the same order you recorded them on your chart. Start on the left with the smaller one, and work to the right.

2. On the "Perimeter" axis, record the perimeters of each of your rectangles. Record them in the same order you recorded them on your chart. Start close to the bottom with the smaller one, and work upwards.

3. On the graph, plot the points for the two rectangles you created. Join the points to create a line.

Vocabulary

line graph: A graph that uses a line or lines to show data or information and its relationships

Technology

Use a spreadsheet application to demonstrate what happens when dimensions (either length or width) are increased by 1 unit. Create an electronic line graph from the data in the spreadsheet. Compare your result with the hand-drawn one in Build Your Understanding.

4. Repeat steps 1 to 3 on the area graph, recording your two area measurements as you increased a dimension.

5. What would your line graphs look like if you continued increasing the original rectangle, either by length or width, by 2 cm, then by 3 cm? Complete your chart and graphs, expanding the pattern up to 4 cm.

Lesson 9: Changing the Dimensions of a Rectangle

What Did You Learn?

Meet in small groups to share, compare, and discuss your rectangles, charts, and graphs.

1. What happened to the perimeter of the rectangle when you increased the dimension by 1 cm? How do you know?

2. What happened to the area of a rectangle when you increased the dimension by 1 cm? How do you know?

3. What would happen to the perimeter and area of a rectangle if you decreased the dimension by 1 cm? Explain. How could you prove your conclusion?

Journal

Describe in your own words what happens to the perimeter of a rectangle when you increase a dimension by 1 cm. Describe in your own words what happens to the area of a rectangle when you increase a dimension by 1 cm.

Practice

Math Problems to Solve

What will happen to the perimeter and area of your original rectangle if you increase two dimensions by 1 cm at the same time?

1. Start with your original rectangle. Predict what will happen to the perimeter and area of the rectangle when you increase both the width and the length by 1 cm. Record your predictions. Calculate the perimeter and area of the new rectangle. How do your results compare with your predictions?

2. Make three more rectangles. For each rectangle, increase both the length and the width by 1 cm more than the previous rectangle. Calculate and record the perimeter and area of each rectangle.

3. Organize your results in a chart like the one you made in Build Your Understanding. Meet with a partner to share, compare, and discuss your chart results.

4. What conclusions can you draw from your experiment?

Lesson 10

Rectangles, Squares, and Diagonals

JOURNEY NOTE

PLAN:
You will draw rectangles on grid paper then create a diagonal and explore relationships between the resulting triangles and the original rectangle.

DESCRIPTION:
Did you ever wonder how the huge stones of the Egyptian pyramids were lifted into place? Experts think that triangle-shaped ramps were used. You can see many examples of right-angle triangles in early civilizations.

Get Started

You will explore some relationships among rectangles, diagonals, and right-angle triangles.

You Will Need
- grid paper
- scissors

1. Draw a rectangle on grid paper.
2. Draw a diagonal to divide the rectangle once. What shapes are formed when you draw the diagonal?
3. Cut out your rectangle. Then cut along the diagonal.
4. Place the shapes you cut out on top of each other. How do the shapes compare?

Vocabulary

congruent: Exactly the same size and shape
congruent triangles: Triangles that are the same size and shape
diagonal: A line segment that joins two opposite vertices on a two-dimensional shape

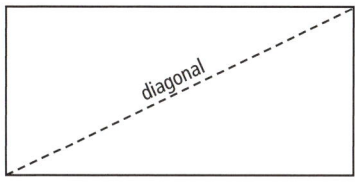

equal angles: Angles that have the same measure
right angle: A square corner angle. A right angle is 90 degrees (90°). On a shape such as a triangle or rectangle, we show the right angle by placing a tiny square where the two line segments that form the right angle meet.

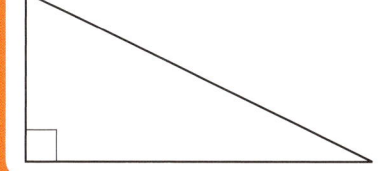

Lesson 10: Rectangles, Squares, and Diagonals

Journal

Make notes about your comparison observations. Note as many features the two shapes share as you can. Use mathematical words such as "congruent," where they apply.

Tip

Use the shapes you cut out to support your answers.

Build Your Understanding

Explore the Relationship Between Right-Angle Triangles and Rectangles and Squares

You Will Need
- geoboard or grid paper

1. Is every right-angle triangle half of a rectangle or a square? Record your prediction. If you do not think so, give an example of when a right-angle triangle is not half of a rectangle.

2. Work with a small group of classmates. On geoboards or grid paper, make a number of right-angle triangles with different dimensions.

3. For each right-angle triangle you create, add a congruent right-angle triangle along the side opposite to the right angle. Then, discuss and record answers to these questions.

   a) What shape(s) did you create when you combined the two identical right-angle triangles?

   b) How does the area of each right-angle triangle compare with the area of the rectangle or square you created when you put the two right-angle triangles together? How do you know?

   c) Does this pattern hold for every right-angle triangle you created? Give evidence to support your conclusions.

4. In your group, find the area of the triangle by counting squares. Then, give another method for finding the area of the right-angle triangle. In your journal, write your method as a rule or formula. Share your rule or formula with the class.

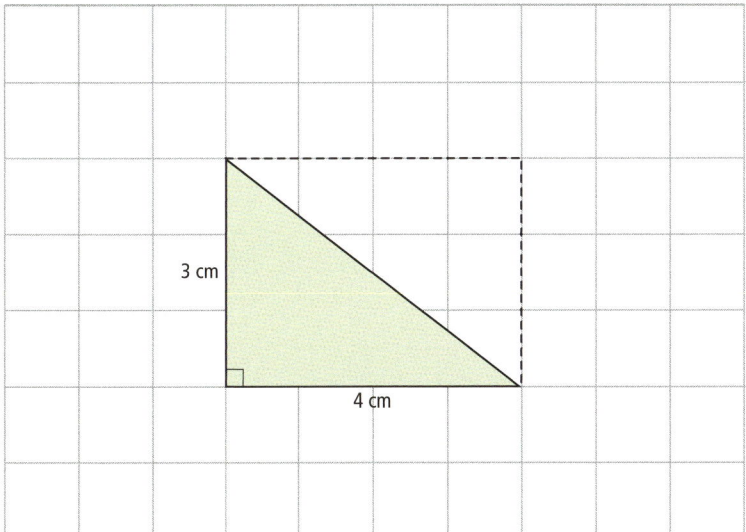

What Did You Learn?

1. Meet with another group to discuss the predictions you made at the beginning of the activity. Is every right-angle triangle half of a rectangle or a square? Give evidence to support your conclusion.

2. What happens when you divide a rectangle or square with a diagonal?

3. How can you use this knowledge to calculate the area of a right-angle triangle?

4. Can you use the same method or formula for finding the area of a right-angle triangle to calculate the area of a different kind of triangle? Explain why or why not.

Technology

In a draw application, use the polygon tool and the copy and rotate features to prove your findings from this lesson's Build Your Understanding. Use the text tool to complete an electronic journal entry explaining your conclusion.

Practice

Thinking back to the Get Started section of this lesson, answer these questions about what happened when you divided your rectangle with diagonals.

1. Were the two new shapes that were created congruent, no matter what the dimensions of the rectangle? Explain.
2. What kind of triangle did you create by making the diagonal? How do you know?
3. What fraction of the original rectangle is each triangle? How do you know?
4. How does the area of each triangle compare with the area of each original rectangle? How do you know?

Show What You Know
Review: Lessons 7 to 10, Measurement

1. Explain how you might find the area of a pair of scissors.
2. Where would you use perimeter and area in real life?
3. Explain how to calculate the area of a right-angle triangle. Use pictures, numbers, and words.

Chapter Review

Chapter 5

MEASUREMENT
GEOMETRY AND SPATIAL SENSE

1. Susan can walk 4.5 km in an hour. How far will she travel in 20 h? Use pictures, numbers, and words.

2. What unit would be used to measure each of the following?
   a) the thickness of your fingernail
   b) the length of the schoolyard
   c) the distance from Saskatoon to Regina
   d) the length of a hockey stick
   e) the width of a hockey stick

3. Tom is 1.43 m tall. How else can you state this height?

4. Complete these conversions:
   a) 1 m = ■ cm or ■ mm or ■ dm
   b) 3 m = ■ dm
   c) 4 m = ■ mm
   d) 800 cm = ■ m
   e) 450 cm = ■ m

5. Calculate the perimeter of each of these shapes:

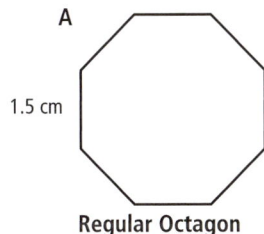

Regular Octagon (1.5 cm)

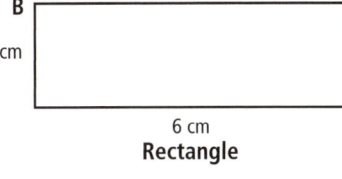

Rectangle (2 cm by 6 cm)

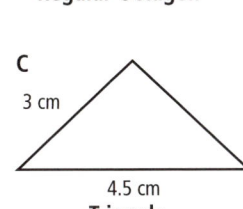

Triangle (3 cm, 4.5 cm)

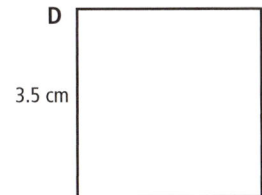

Square (3.5 cm)

6. If the perimeter of a triangle is 12 cm and two of its sides are each 4 cm long, what is the length of the third side? What type of triangle is this?

Chapter Review 257

7. Explain how to find the perimeter of this shape:

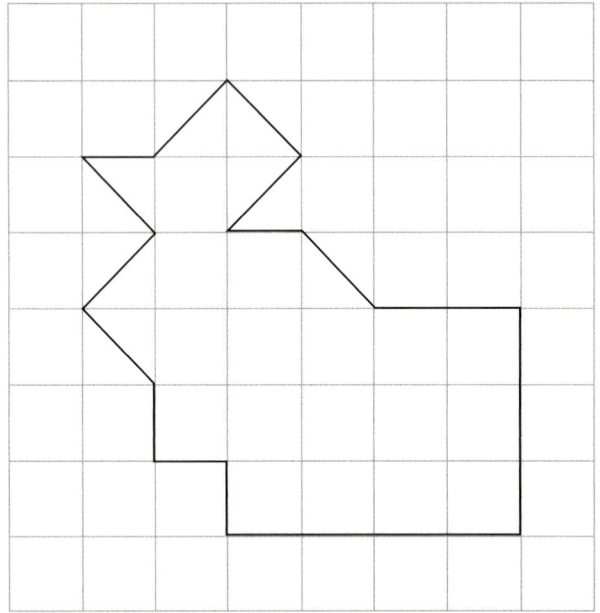

8. Estimate the area of the shape in question 7.

9. Use grid paper to make three irregular shapes that are 30 cm².

10. Find the area and perimeter of these rectangles:

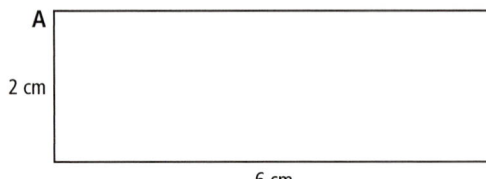

11. Look at the dimensions of this shape.
   a) What is the perimeter?
   b) What is the area?

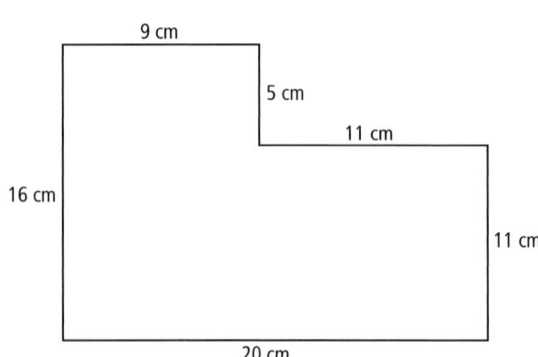

258   Chapter 5: Math in Ancient Cities

12. How many different rectangles can have a perimeter of 20 cm? How many different rectangles are possible? You should use full units. Give the dimensions of each rectangle. Use grid paper or a geoboard to show your rectangles.

13. Find the area of each triangle:

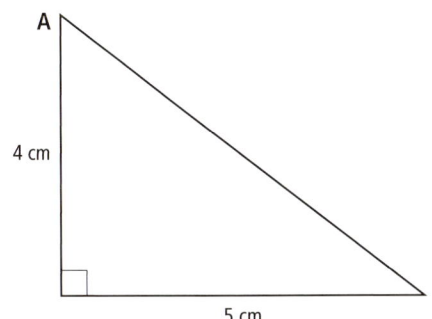
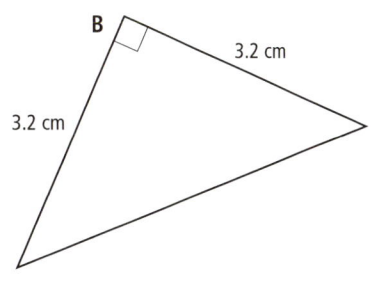

14. Explain how you calculated the area of each triangle in question 13.

15. Use a story-maker application (for example, Storybook Weaver, EasyBook) to create a mystery story involving measurement. Perhaps clues requiring measuring could help solve the mystery.

Chapter Wrap-Up

Chapter 5

MEASUREMENT GEOMETRY AND SPATIAL SENSE

You have reached the end of Chapter 5. Throughout this chapter, you have learned about many interesting and important cities from different early civilizations. You have learned key mathematical concepts: metric distance measurement units; the relationships between these measurement units; and how to estimate and calculate the perimeter and area of squares, rectangles, triangles, and irregular shapes.

The following project will give you an opportunity to apply and develop your mathematical skills and knowledge.

Imagine that you are part of a team of archeologists and you have discovered the ruins of an important city from an early civilization.

- There are four structures on the site.
- The first structure is a rectangle.
- The second structure is also a rectangle. The area of the second structure is half the area of the first structure.
- The third structure is the shape of a different regular polygon.
- The final structure has an irregular shape.

Chapter 5: Math in Ancient Cities

1. Draw a map of the archeological site showing the four different structures. You might want to plan your map before you draw it. Label the structures and explain how they were used on the back of your map or on a separate piece of paper.

2. On the map, give the dimensions, perimeter, and area of each structure. Give dimensions that would be reasonable for this type of structure. For example, if it is a temple, the structure would be quite large. Use measurement units appropriate to structures of this size.

3. Choose an early civilization that interests you. Add features to the structures that would be found in this early civilization. For example, if you selected a Northwest Coast native culture, one building might be a longhouse, and a feature of this structure would be house poles. Research important shapes in the culture, such as the shapes of temples, palaces, or important monuments.

4. To make a complete, accurate map, include these things:
   - the correct shapes on your map
   - proper labelling of structures
   - an explanation of how the structures were used
   - accurate labelling of dimensions
   - appropriate measurement units
   - a scale
   - correct calculation of perimeter and area

   Share your site map with classmates.

Chapter 6

Math in Monuments and Marvels

As you continue exploring the wonders of the past, you will focus on the monuments of ancient cultures and what they tell us about the people who created them. You will see and learn about the phenomenal size of these monuments, their patterned shapes, and the appealing designs that decorate them.

In this chapter, you will identify, describe, measure, compare, and classify two-dimensional shapes and three-dimensional figures. You will use games with number cubes and puzzles such as tangrams to both solve and pose challenging problems.

At the end of this chapter, you will apply the math skills you have learned to create a monument of your own.

The monuments below are all several metres tall. Which mathematical concepts do you think their creators used to design and build them? What patterns do you see? Think of large structures in your area. How was math used to build them? What math concepts do you think of when you look at these structures?

Good luck on the final leg of your journey!

Easter Island statues

Persepolis, Iran

Stonehenge, England

MEASUREMENT

Lesson 1

Estimating and Measuring Heights

JOURNEY NOTE

PLAN:
You will estimate the height of an inukshuk. As well, you will use paper clips and a metre stick to obtain an approximate measurement of the height of your school.

DESCRIPTION:
An inukshuk is a human figure made out of stones. In early times, Inuit people used it to scare the caribou toward hunters. In modern times, the inukshuk is used as a marker to guide travellers.

Inukshuk

Get Started

Imagine that during your travels in the far North you saw an inukshuk with a person standing beside it. How would you estimate the inukshuk's height so that you could describe its size to friends back home?

1. Meet with a classmate to develop a strategy for estimating the height of an inukshuk like the one shown on the right.

2. Record your estimate.

3. Meet with another pair of students to compare and discuss your estimates and estimation strategies. How could you determine which estimate is most accurate?

Journal
Make notes of why, where, and when you must estimate height in your daily life. What strategy or strategies do you use to make these estimates?

Lesson 1: Estimating and Measuring Heights 263

Build Your Understanding

Estimate the Height of Tall Structures

You Will Need
- 10 paper clips per group
- metre stick

How could you accurately estimate the height of your school?

1. Work in a small group. One group member stands beside the school holding a metre stick a little way from the base of the school wall.

2. Another group member holds a paper clip upright at arm's length. He or she walks carefully backwards until the paper clip appears to be the same length as the metre stick.

3. The student with the paper clip remains in the same spot. A third group member stands beside this student, links paper clips together, and holds the resulting paper-clip chain upright. The third group member continues adding paper clips until the length of paper clips appears to be the same height as the school.

4. Count the number of paper clips.

5. Multiply the number of paper clips by one metre to obtain an estimate in metres of your school's height. Write this as a scale.

6. If possible, obtain an accurate measurement of the height of the school. How does the actual height compare with your estimate?

Vocabulary

height: The distance from the base of something to its top; how tall something is
length: The distance from one end of something to the other end; how long something is

Journal

Comment on how you could improve your height estimation strategy.

What Did You Learn?

1. Meet as a class to share, compare, and discuss your school height estimates and estimation strategies.

2. Explain the method you used to estimate the height of the inukshuk. What are the advantages and disadvantages of this method?

3. Explain the method you used to estimate the height of the school. What are the advantages and disadvantages of this method?

4. How could you improve each of these estimation strategies?

Practice

Extension

1. Here are the heights of important monuments and structures from early civilizations.

Structure, Location, Height

| The Great Pyramid of Cheops, Giza, Egypt 147 m | Totem Pole, Victoria, BC, Canada 38.8 m | Stonehenge, Salisbury, England 6.8 m | The Giant Heads, Easter Island, Chile 9.8 m | Borobudur Temple, Java, Indonesia 34.5 m |

a) What heights surprised you? Why?

b) Which is the tallest monument? Which is the shortest?

c) What monuments and structures from early civilizations do you think should be added to the chart? Why? What are their heights?

2. With your classmates, create a bulletin-board display showing pictures or drawings of important monuments from early civilizations. Give the height of each monument in an appropriate unit of measurement.

3. Use the format of the chart to make your own chart of monuments' heights in your community. Apply estimation strategies you learned in this lesson to find approximate heights of local monuments.

Technology

Use the Internet to research Inuit inukshuks. Find the height of as many of these structures as you can. Share your findings with the class. Ask your teacher for permission to use the Internet.

Tip

War memorials are common monuments in many communities. Another structure you can measure is the school's flag pole.

4. Research one of the monuments or structures shown on the previous page. Find out how the monument or structure was built. Share your findings with the class.

5. Use an estimation strategy you learned in this lesson to find the approximate height of your house or apartment building. Meet as a class and share your results. Use these results to create a class comparison chart.

GEOMETRY AND SPATIAL SENSE
PATTERNING AND ALGEBRA

Lesson 2

Comparing Pyramids and Prisms

JOURNEY NOTE

PLAN:
You will review your understanding of pyramids and prisms. Then you will make models of different types of pyramids and prisms, and look for patterns in the number of faces, edges, and vertices in these solids.

DESCRIPTION:
Many structures of the past were built in the shape of pyramids and prisms, like El Castillo in Mexico and the homes of the Pueblo in the southern United States.

El Castillo, Mexico

Pueblo home, Taos, New Mexico

Get Started

A pyramid is a solid figure with a polygon base and triangular faces that meet at one common point.

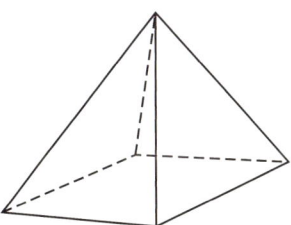

In a prism, there are two faces that are congruent and parallel. The other faces are rectangular.

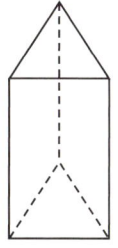

Vocabulary

base: A surface on which a figure can stand
edge: The line segment where two faces of a figure meet
face: The flat side of a figure
prism: A three-dimensional figure with two faces that are congruent and parallel and other faces that are parallelograms
pyramid: A solid figure with a polygon base and triangular faces that meet at a common point
solid figure: A three-dimensional shape

Lesson 2: Comparing Pyramids and Prisms

Pyramids and prisms are often named for their bases. For example, the pyramid on the previous page is a square-based pyramid. The prism below it is a triangular prism, and the prism below here is a rectangular prism.

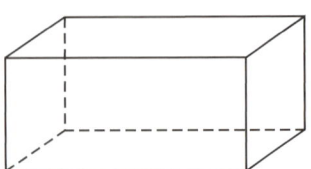

Meet with a partner. Which of the monuments shown on page 265 is a pyramid? Which is a prism? What kind of pyramid or prism is each? Explain how you know.

Journal

Write vocabulary definitions in your own words. Create a picture to help you understand each definition.

Build Your Understanding

Make and Classify Pyramids and Prisms

You Will Need
- modelling clay
- toothpicks or pieces of drinking straw
- sticky notes

1. Work in a group and use the modelling clay and toothpicks or pieces of straw to make pyramids and prisms with each of the bases below.

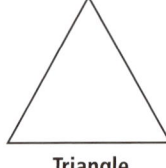  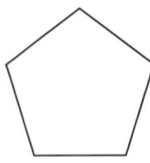 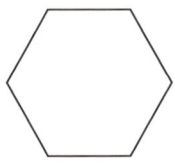

Triangle   Square   Pentagon   Hexagon

2. On sticky notes, name each pyramid and prism you constructed. For example, a prism with a triangular base is a triangular prism.

Tip

To save time, divide model-making responsibilities equally among group members.

Chapter 6: Math in Monuments and Marvels

3. In your notebook, create charts like these. Use the pyramids and prisms you made to complete each chart.

Pyramids				
Base	Number of Faces	Number of Edges	Number of Vertices	Sketch
Triangle				
Square				
Pentagon				
Hexagon				

Prisms				
Base	Number of Faces	Number of Edges	Number of Vertices	Sketch
Triangle				
Square				
Pentagon				
Hexagon				

What Did You Learn?

In your group, discuss and record the answers to the following.

1. What patterns do you notice in the information you collected about pyramids? What is similar and different among the pyramids you explored?

2. What patterns do you notice in the information you collected about prisms? What is similar and different among the prisms you explored?

3. Create group definitions of pyramid and prism. Make labelled pictures to support your definitions.

Journal
Summarize what you've learned about pyramids and prisms in this lesson.

Technology
Conduct an Internet search for architectural photographs. Copy and paste them into a paint application. Create a collage of pictures containing a variety of pyramids and prisms. Use paint tools to outline the prisms and pyramids.

Lesson 2: Comparing Pyramids and Prisms

Practice

1. **a)** Look for pyramids and prisms in your classroom, school, and community.

   b) Make a list of all the pyramids and prisms that you find.

   c) Beside each item on your list, write what kind of pyramid or prism it is and explain how you know, using math vocabulary.

Extension

2. **a)** Look for pyramids and prisms in photographs of the monuments of early civilizations.

   b) Mount each picture on cardboard.

   c) Write captions under each one. Each caption should
   - tell whether the monument shape is a pyramid or a prism
   - tell what kind of pyramid or prism it is and explain why, using important mathematical words from the lesson
   - give information about where and when the people who made the monument lived

   d) You can work with a large group to organize your photographs into a bulletin-board display.

   e) Classify your pictures and caption information by shape or by early civilization.

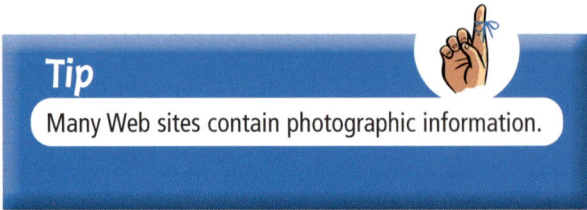
Tip
Many Web sites contain photographic information.

GEOMETRY AND SPATIAL SENSE

Lesson 3

Pyramids, Prisms, and Nets

JOURNEY NOTE

PLAN:
You will construct models of square-based pyramids and rectangular prisms. Then, you will create a variety of nets for square-based pyramids and rectangular prisms.

DESCRIPTION:
Imagine you wanted to create a model of an ancient village. You could only use flat, paper cut-outs to make the three-dimensional figures such as pyramids and prisms in village buildings. How could you do it?

How many pyramids and prisms can you find in this picture?

Get Started

Work with a classmate to find an example of a square-based pyramid in the village picture. How many faces does the pyramid have? Remember to include the base. Draw all the faces on a piece of scrap paper.

Next, find an example of a rectangular prism in the village picture. How many faces does the rectangular prism have? Draw each face on a piece of scrap paper. Remember that some faces will be congruent and common edges will be equal.

Lesson 3: Pyramids, Prisms, and Nets

You Will Need
- scissors
- scrap paper
- adhesive tape
- square-based pyramid and rectangular prism nets on reproducible page

> **Vocabulary**
>
> **net:** A pattern that can be folded to make a three-dimensional figure

1. Work with a classmate. Cut out the square and triangular faces needed to construct a model of a square-based pyramid.

2. Use adhesive tape to fasten each face to the square base. Then, make sure all triangular faces meet at a vertex. Fasten the faces together to form the pyramid.

3. Cut out the four rectangles and two squares needed to construct a model of a rectangular prism.

4. Use adhesive tape to hold the faces and bases in place.

   If you opened up your three-dimensional models to make a flat, or two-dimensional shape, you would have a net.

5. Cut out the nets from the reproducible page to create another square-based pyramid and rectangular prism. Use adhesive tape to hold the faces and bases in place. With the class, discuss which set of models you found easier to construct. Explain why.

Build Your Understanding

Draw and Explore Nets

1. Work with a partner. Use your model pyramid to help you make drawings of as many different nets as you can for a square-based pyramid. Record net sketches in your notebook.

> **Tip**
>
> Meet with other pairs of students to share and discuss ideas for net possibilities. You may need to test any nets you are unsure of.

2. Next, use your model prism to help you make drawings of as many different nets as you can for a rectangular prism. Record net sketches in your notebook.

Chapter 6: Math in Monuments and Marvels

What Did You Learn?

As a class, share, compare, and discuss all the different nets that were made for the pyramid and prism.

1. How many nets did you find for this squared-based pyramid? Refer to sketches to support your answer.

2. How many nets did you find for the rectangular prism? Refer to sketches to support your answer.

3. Will all rectangular prisms have the same number of nets? Explain why or why not. Use sketches and models to support your answer.

Practice

1. Will the net to the right make a pyramid?

   To prove your answer, trace the net, cut it out, and try to fold it to construct a pyramid.

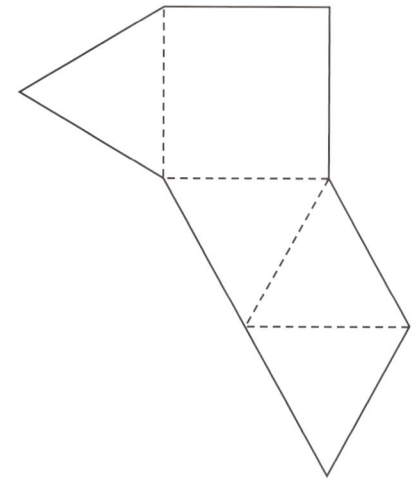

2. Will the net to the right make a rectangular prism?

   To prove your answer, trace the net, cut it out, and try to fold it to construct a prism.

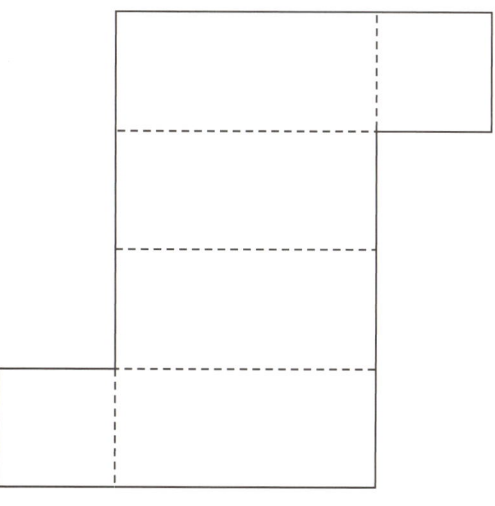

Lesson 3: Pyramids, Prisms, and Nets

3. A cube is a rectangular prism with six congruent square faces. Which of the six nets below will form a cube? Explain why.

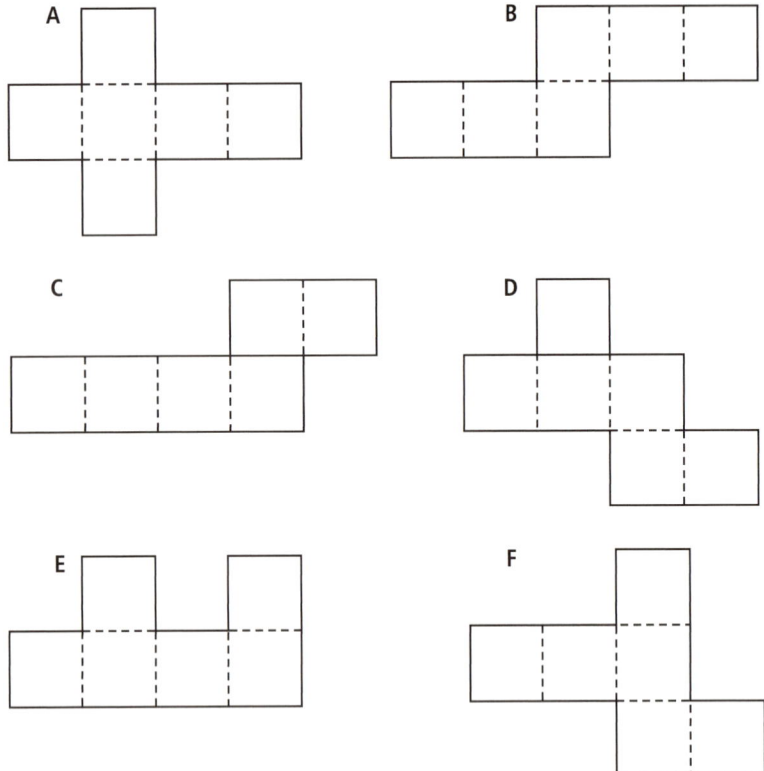

4. a) Find a box that is a rectangular prism and that you have permission to cut.

   b) In your notebook, draw what you think the net will look like if you cut the box to make a flat, two-dimensional shape.

   c) Cut the box along its edges, making sure that all the faces remain connected.

   d) Draw a picture of what the box looks like flat.

   e) How do your two net drawings compare?

   f) Are they the same or different? Explain why they might be different.

Technology

Use a geometry application such as tabs+ to create pyramids and prisms with different bases. Rotate the objects to view all their faces. In a chart, record the number of faces, edges, and vertices. Print out nets for your pyramids and prisms. Then, assemble them.

Vocabulary

cube: A three-dimensional solid with six congruent square faces

Chapter 6: Math in Monuments and Marvels

GEOMETRY AND SPATIAL SENSE

Lesson 4
Exploring the Faces of Pyramids and Prisms

JOURNEY NOTE

PLAN:
You will learn about pyramids and prisms that have triangle, pentagon, hexagon, and octagon faces or bases. Then, you will use these two-dimensional shapes to draw three-dimensional objects on grid paper.

DESCRIPTION:
The bases or faces of pyramids and prisms are not always squares or rectangles. You can see this in buildings and monuments throughout history.

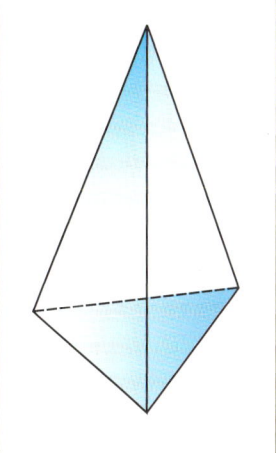
This pyramid has a triangular base.

The turret on this castle is an octagonal prism (eight-sided).

Get Started

In Lesson 3, you spent time exploring, drawing, and making models of square-based pyramids and rectangular prisms. Can pyramids and prisms have bases other than squares or rectangles?

Different types of pyramids and prisms can be constructed by using different-shaped polygons for their bases. These types of pyramids and prisms are named for the polygons that form their bases. The polygons on this page can be used to create pyramids and prisms.

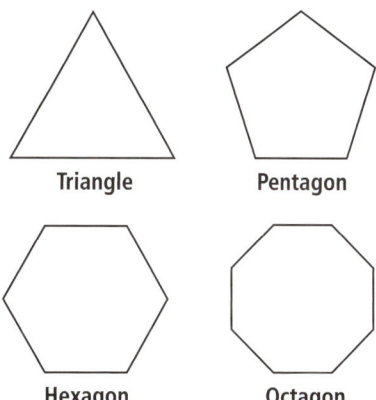

Triangle   Pentagon

Hexagon   Octagon

Tip

If you are unsure about the meaning of any mathematical word, look it up in the glossary at the end of this book. If you can't find it there, try a mathematical dictionary.

Lesson 4: Exploring the Faces of Pyramids and Prisms

Vocabulary

heptagon: A polygon with seven sides; hepta means seven
hexagon: A polygon with six sides; hexa means six
nonagon: A polygon with nine sides; nona means nine
octagon: A polygon with eight sides; octa means eight
pentagon: A polygon with five sides; penta means five
polygon: A closed shape that is formed by three or more line segments
solid: A three-dimensional figure

Journal

Make a list of polygon definitions. Draw and label pictures to support your definitions. Underline in red the prefix that tells you how many sides each polygon has.

Build Your Understanding

Draw and Compare Pyramids and Prisms

You Will Need
- grid paper
- ruler
- pattern blocks

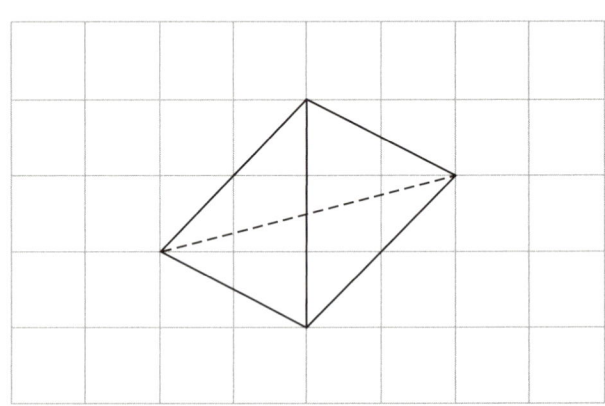

1. Work with a partner to examine the picture. What is the name of the solid? How do you know?

2. Draw a triangle on a piece of grid paper. Make the triangle the same size as the base of the pyramid here. Use the triangle to draw a triangular prism. How did you do this?

3. Compare your triangular prism with the triangle-based pyramid on this page. How are they similar? How are they different?

Booklink

The X-ray Picture Book of Big Buildings of the Ancient World by Joanne Jessop (Viking: Toronto, ON, 1994). This book describes and explores 10 of the ancient world's most extraordinary buildings.

4. On grid paper, draw a pentagon, a hexagon, and an octagon. Use the two-dimensional shapes to draw pyramids with these polygons as their base.

5. On grid paper, draw a pentagon, a hexagon, and an octagon. Use these two-dimensional shapes to draw prisms with pairs of these polygons as their parallel faces.

Journal

Make note of similarities and differences you observe between and among the polygons and solids.

What Did You Learn?

Meet with classmates to share, compare, and discuss your drawings.

1. What did you find most challenging about drawing pyramids and prisms?

2. What tips would you give a classmate who was having problems drawing pyramids and prisms?

3. When you draw a prism and a pyramid with the same shape of base or face, what is the main difference between the two types of solids?

Practice

1. Name and describe the following solids using mathematical words:

A
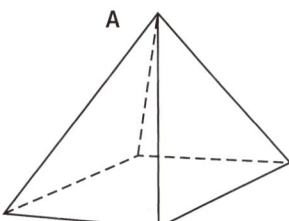

B
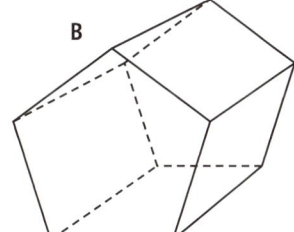

C
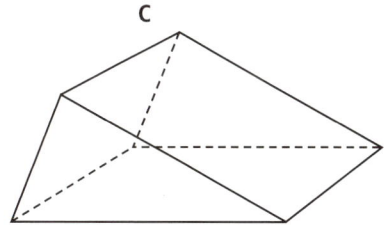

Lesson 4: Exploring the Faces of Pyramids and Prisms

2. Use the pentagon to draw a pentagon-based pyramid. Use the octagon to draw an octagonal prism. Label your drawings. Look at the illustrations in question 1 for ideas on how to draw your pyramid and prism.

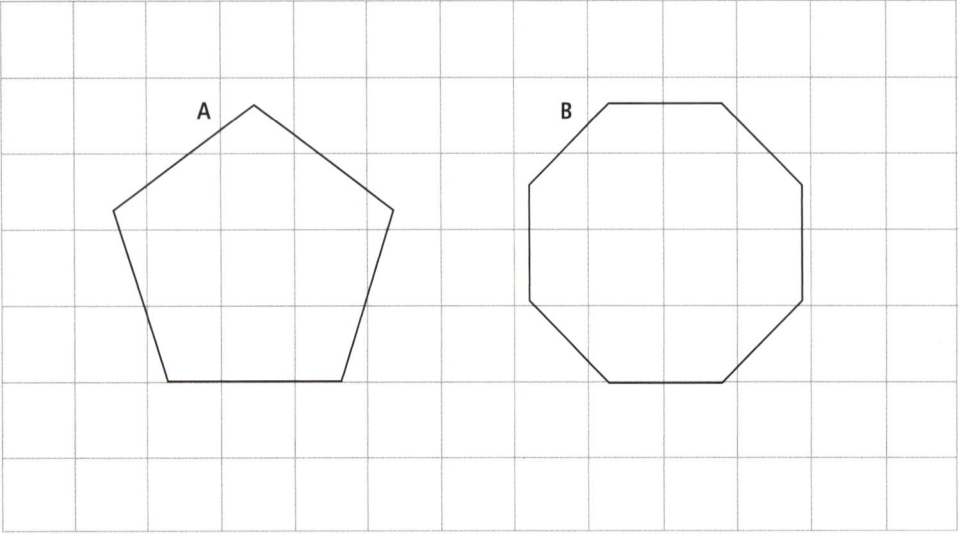

3. What are the similarities and differences between a triangle-based pyramid and a triangular prism? Use pictures, numbers, and words.

4. What are the similarities and differences between an octagon-based pyramid and a hexagonal prism? Use pictures, numbers, and words.

5. Draw two different rectangular prisms using the same faces. Label the dimensions of both prisms.

Technology

Use a puzzle-making application or visit www.puzzlemaker.com. Create a crossword puzzle using mathematical terms and phrases related to the study of pyramids and prisms. Challenge a classmate to complete your puzzle.

Chapter 6: Math in Monuments and Marvels

GEOMETRY AND SPATIAL SENSE
MEASUREMENT
NUMBER SENSE AND NUMERATION

Lesson 5

Creating and Solving Problems About the Great Pyramid

JOURNEY NOTE

PLAN:
You will learn measurement facts about the Great Pyramid of Cheops then use these facts to create and solve problems about measurement, geometry, and operations with decimals.

DESCRIPTION:
The people of ancient Egypt built pyramids to house the bodies of their dead kings. The pyramids are the oldest and biggest stone structures in the world. Of the Seven Wonders of the Ancient World, the three largest pyramids at Giza form the only remaining one.

The Great Pyramid of Cheops

Get Started

Here are some facts about the Great Pyramid of Cheops.

1. Experts think the pyramid was finished around 2560 B.C.

2. The original height of the Great Pyramid was 146.6 m. However, because of erosion and other time factors, it now stands at 138.75 m.

3. Each side of the Great Pyramid faces one of the main points on a compass: north, south, east, and west.

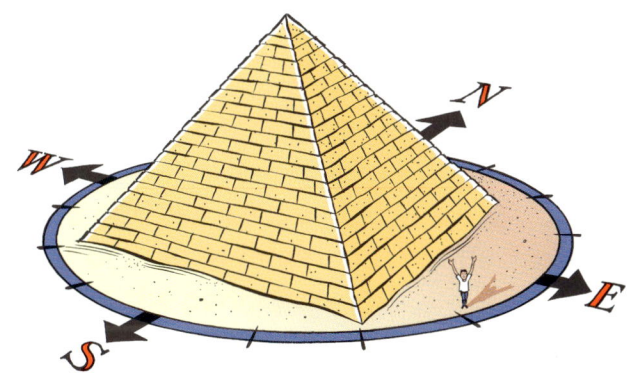

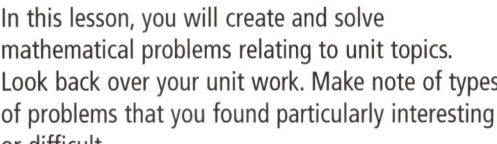

Journal

In this lesson, you will create and solve mathematical problems relating to unit topics. Look back over your unit work. Make note of types of problems that you found particularly interesting or difficult.

4. The Great Pyramid is a square-based pyramid. At its base, each side has a length of 230.37 m.

5. The four edges of the triangles rise at an angle of a little more than 51° from the base.

6. For 43 centuries, the Great Pyramid was the world's tallest structure.

7. The pyramid is made of more than 2 million stone blocks.

8. Each block in the pyramid weighs more than 1800 kg.

Work in a small group to create your own mathematical problems using information about the Great Pyramid. Here are some model problems to get you started:

- How does the height of the Great Pyramid compare with the height of the CN Tower, or the height of a mountain or monument in your area? How could you show this comparison in a drawing?

- How much of your schoolyard and the surrounding neighbourhood would the base of the Great Pyramid cover? Explain using pictures, numbers, and words.

Build Your Understanding

Pose and Solve Pyramid Problems

1. Work in a small group to create five problems relating to facts about the Great Pyramid of Cheops.

    a) Write a problem to cover each of the following unit topics:
    - measurement
    - calculating perimeter and area
    - operations (adding, subtracting, multiplying, dividing) with decimal numbers
    - naming angles
    - an important unit topic of your choice

    b) After writing a rough draft of each problem, write a good copy on a piece of cardboard. On the back, provide the answer and any formulas or strategies needed to solve the problem.

2. Create some comparisons of features of the Great Pyramid to things you are familiar with. Good writers often use comparisons to make ideas clear to their readers. For example, "Each stone weighed more than two elephants."

3. a) List all Great Pyramid comparisons the group created.

   b) Use your comparisons to write a paragraph caption to go with the picture of the Great Pyramid. In your caption, give readers a sense of the pyramid's size and the great accomplishment of early Egyptians who designed and built it.

Tip

Look back at your journal notes on problems you found interesting or particularly challenging. Use the amazing facts about the Great Pyramid to produce these types of problems.

Booklink

An Egyptian Pyramid by Jacqueline Morley (Simon and Schuster: New York, USA, 1991). Text and illustrations show the construction and uses of the pyramids in ancient Egypt and explore how people lived and worked in that time.

What Did You Learn?

1. Challenge another group to solve problems your group created. Share and discuss problem-solving strategies.

2. Share your paragraph captions with the class. Explain why you made the comparisons you did. Also, describe any calculations you made.

Practice

Extension

There are two other pyramids that stand beside Cheops: Khafre, which is 136.4 m high, and Menkaure, which is 66.5 m high.

1. Research these pyramids to find information about their sizes, shapes, and histories.

2. Use this information to create mathematical problems or interesting measurement comparisons.

3. Share your problems and comparisons with classmates.

Lesson 6
Constructing Solids Made From Cubes

JOURNEY NOTE

PLAN:
You will estimate, then count, the number of cubes needed to construct three-dimensional figures. Then, you will construct model solids that match pictures of the solids and check your estimates.

DESCRIPTION:
The very first pyramids the Egyptians built — even before the Great Pyramid of Cheops — were step pyramids. These early pyramids were made of huge limestone blocks. The Step Pyramid at Saqqara, made for King Djoser around 2650 B.C., was the first monumental stone building ever constructed in the world.

The Step Pyramid at Saqqara, Egypt

Get Started

You might remember that a cube is a three-dimensional solid with six congruent square faces. The solids here are made from cubes.

A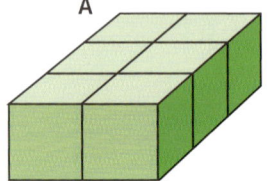

B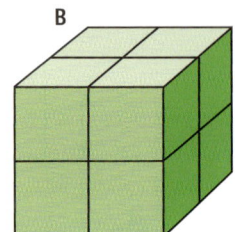

You Will Need
- 14 cube blocks

1. Work with a partner. Name the solid shown in each picture to the right. How can you tell what type of solid it is?

2. How many cubes do you estimate are in each solid? Use your blocks to make the shapes. How many blocks did you use to make each solid?

Vocabulary

cube: A three-dimensional solid with six congruent square faces
volume: The amount of space an object takes up. A measurement unit of volume is the cubic centimetre, or cm^3.

Chapter 6: Math in Monuments and Marvels

Build Your Understanding

Estimate the Number of Cube Blocks in an Object

You Will Need
- 30 cube blocks per group

You will build pyramid-shaped structures with steps from cube blocks. Work in a small group.

1.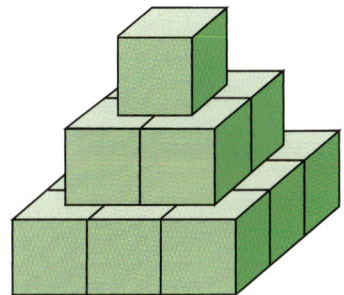
   a) Estimate how many cubes there are.
   b) Record your estimate and explain how you arrived at the estimate.
   c) Build the step pyramid with the group's blocks.
   d) Record how many blocks your group used.
   e) How does your estimate compare to the actual number of blocks used?

2.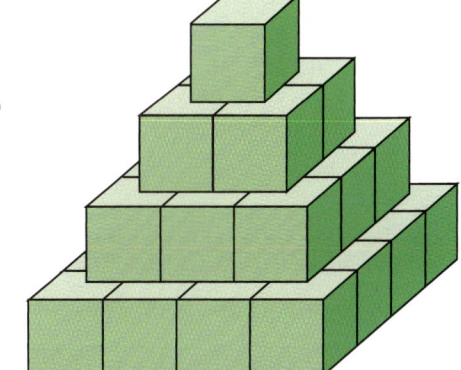
   a) Estimate how many cubes there are.
   b) Record your estimate and explain how you arrived at the estimate.
   c) Build the step pyramid with the group's blocks.
   d) Record how many blocks your group used.
   e) How does your estimate compare to the actual number of blocks used?

What Did You Learn?

1. What pattern(s) did you notice in the number of cubes needed to build a pyramid when the number of cubes along the side of the base was increased by one?

2. a) Predict the total number of cubes in a step pyramid if you have 5 cubes along the side of the pyramid base.
   b) Explain your reasoning.
   c) Use the cube blocks to build this pyramid and test your prediction.

Lesson 6: Constructing Solids Made From Cubes

Practice

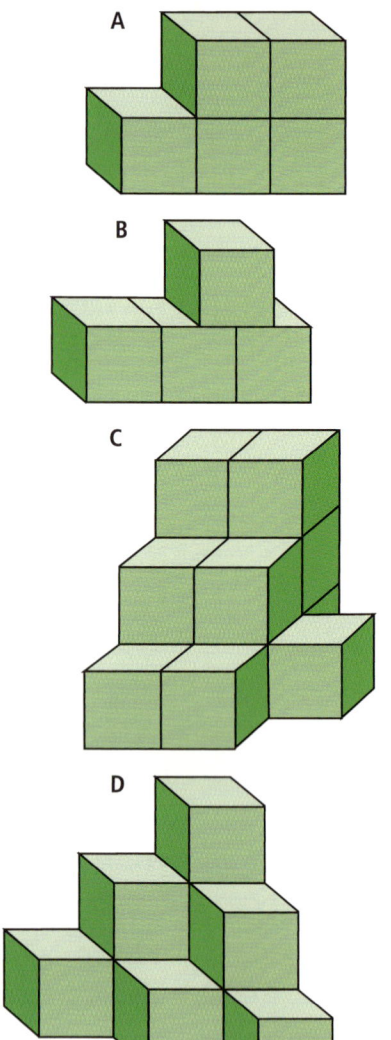

1. **a)** Estimate the number of cubes used to make solids A to D and record your estimate.
   b) Use your blocks to make each solid.
   c) Record the number of blocks you used to make each solid.
   d) Which solid did you find most challenging to make? Why?
   e) How did your estimates compare to the actual number of blocks used for each solid?

2. **a)** Predict how many cubes there would be in a pyramid with 6 cubes along one side of the base.
   b) Explain why you think so.
   c) Build the pyramid to test your prediction.
   d) Predict the number of cubes in a pyramid with 7 cubes along one side of its base.
   e) Explain why you think so.
   f) Build the pyramid to test your prediction.

3. Describe a way of determining the number of cubes without building the pyramid.

Show What You Know

Review: Lessons 1 to 6, Measuring and Pyramids, Prisms, and Solid Figures

1. **a)** Choose three trees on school grounds.
   b) Using what you have learned in this chapter, estimate their heights.
   c) Compare your estimates with those of your classmates.

2. **a)** Use materials to build different types of pyramids and prisms.
   b) Draw and label your constructions.

3. **a)** Make a solid using cubes.
   b) Have a partner estimate how many cubes there are in your solid figure.
   b) Then, have your partner use cubes to make the solid.

Chapter 6: Math in Monuments and Marvels

GEOMETRY AND SPATIAL SENSE

Lesson 7

The Tangram Puzzle

JOURNEY NOTE

PLAN:
You will learn how to flip, turn, and move tangram pieces in many ways to create polygons and animal creatures. You will also use mathematical language to explain how you created these figures.

DESCRIPTION:
A tangram is a very old Chinese puzzle. According to legend, one day a wise old man dropped his favourite square tile. It broke into seven pieces. He looked at the pieces on the floor for a long time, and they gave him the idea of making a seven-piece puzzle.

A tangram puzzle

Get Started

You Will Need
- scissors
- tangram reproducible page

1. Work with a classmate.
   a) Examine all the pieces in the tangram puzzle.
   b) Name all the polygons in the puzzle.
   c) What other mathematical words can you use to describe the shapes and their relationships?

2. a) Cut out the tangram puzzle.
   b) Separate the two sets of congruent triangles in the puzzle.
   c) Predict what shape you will create when you put each pair of triangles together.
   d) Record the prediction in your notebook.
   e) Combine the triangles.
   f) What shapes did you create?

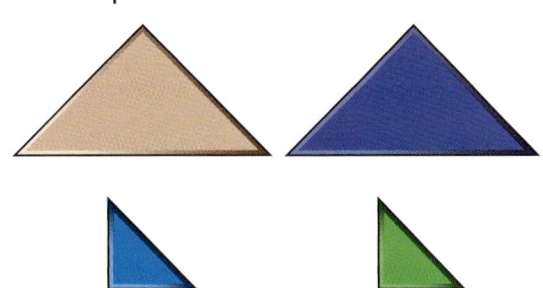

Lesson 7: The Tangram Puzzle 285

3. a) Use three tangram shapes to make a rectangle.

   b) In your notebook, draw the rectangle you created.

4. a) Use four tangram shapes to make a square.

   b) Draw the square in your notebook, showing the shapes you used to create it.

5. a) Use five puzzle pieces to make a different rectangle.

   b) Draw the rectangle in your notebook, showing the shapes you used to create it.

Vocabulary

parallelogram: A quadrilateral whose opposite sides are parallel

tangram: An ancient Chinese puzzle made from a square cut into seven pieces: two large triangles, one medium-sized triangle, two small triangles, one square, and one parallelogram

Build Your Understanding

Solve Tangram Problems

You Will Need
- tangram puzzle

1. Work with a partner. Experiment with your tangram puzzle pieces to see if you can create the following:
   - square using only one tangram piece
   - square using two tangram pieces
   - square using three tangram pieces
   - square using five tangram pieces
   - square using six tangram pieces
   - square using all seven tangram pieces

   In your notebook, record all of the squares you can make.

2. In ancient China, tangrams were used to create pictures of animals or flowers to accompany a story.

   a) Use the shapes to make each of the animal figures below.

   b) Draw each animal figure in your notebook, showing the shapes you used to create them.

Booklink

Grandfather Tang's Story: A Tale Told With Tangrams by Ann Tompert (Crown: New York, USA, 1990). Grandfather tells a story about shape-changing fox fairies who try to outdo each other until a hunter brings danger to both of them.

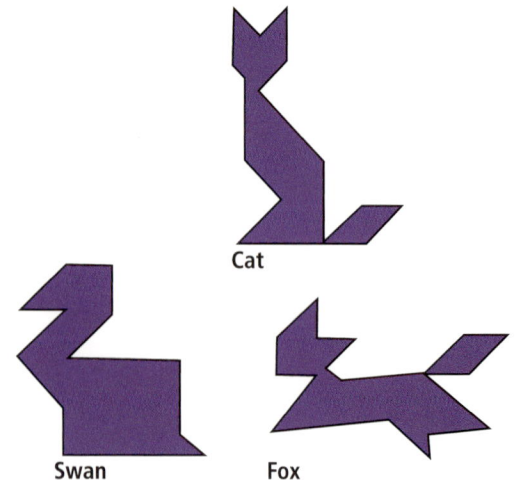

Cat

Swan

Fox

Chapter 6: Math in Monuments and Marvels

3. Explain to another pair of students what tangram polygons you used and how you arranged and moved those shapes to create one of the creatures.

Tip

Experiment with flipping, turning, or moving pieces in new ways to create new shapes.

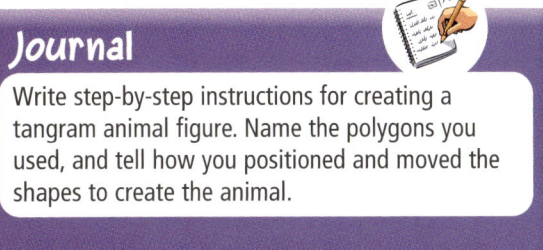

Journal

Write step-by-step instructions for creating a tangram animal figure. Name the polygons you used, and tell how you positioned and moved the shapes to create the animal.

What Did You Learn?

1. Of the shapes you created in Build Your Understanding, which were regular? Which were irregular? Explain how you know.
2. Which strategies did you find most useful when trying to solve tangram problems? Why did you find them useful?

Practice

1. a) Use tangram pieces to create the following creatures.
   b) Record the methods in your notebook, showing the shapes you used to create them.

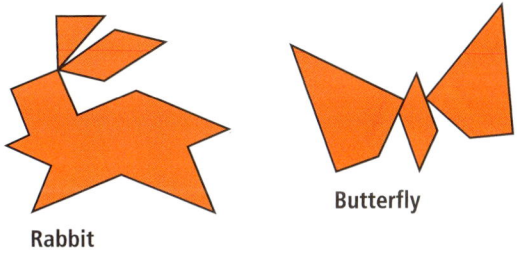

Rabbit

Butterfly

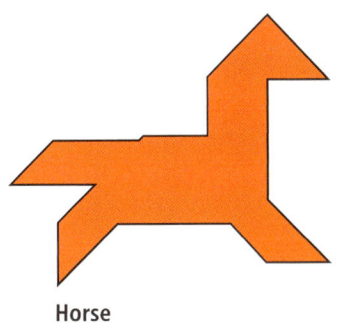

Horse

2. Use tangram shapes to create or support a story about early Chinese civilization. You might use creatures from this lesson or create new ones. Share your tangram story with the class.

Technology

Using a paint application, create pieces for a tangram puzzle. Slide, flip, and turn the pieces to create the animals in this lesson.

Tip

Review your journal, notebook, and the glossary at the end of this book to find mathematical words that describe the polygons.

Lesson 7: The Tangram Puzzle

Lesson 8

Exploring Tangrams Further

JOURNEY NOTE

PLAN:
You will describe and compare polygons found in tangram puzzle pieces. Then you will use tangrams to create puzzle problems for classmates and solve problems created by classmates.

DESCRIPTION:
Early Chinese civilization gave the world many amazing advances in mathematics and technology. For example, the early Chinese produced one of the first books using a block printing method. It consisted of one rolled page that was 4.5 m long!

The Diamond Sutra is the earliest dated book. It was printed in A.D. 868.

Get Started

Let's explore the relationships between the sizes of triangles in tangram pieces.

 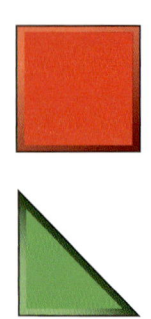  

1. a) What fraction of the small square is the small triangle?
   b) How do you know?
   c) Draw a picture to show your answer.
   d) What mathematical word describes the relationship between the two small triangles?

2. a) What fraction of the big triangle is the medium-sized triangle?
   b) How do you know?
   c) Draw a picture to show your answer.

 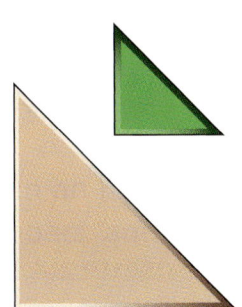 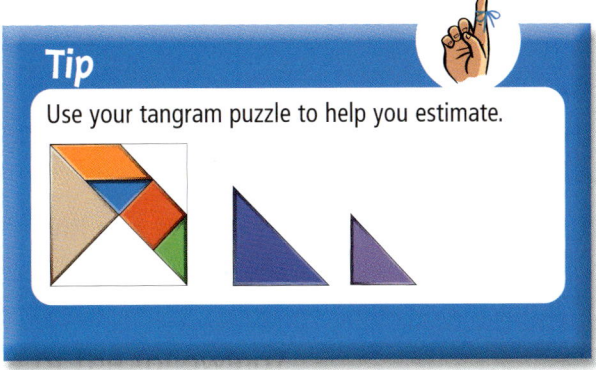

3. a) What fraction of the medium-sized triangle is the smallest triangle?
   b) How do you know?
   c) Draw a picture to show your answer.

4. a) What fraction of the largest triangle is the smallest triangle?
   b) How do you know?
   c) Draw a picture to show your answer.

5. a) Use measurement tools such as protractors and rulers to measure each of the triangles.
   b) Calculate perimeters and areas of the triangles.
   c) Show your work.

Journal

Write a few sentences summarizing the relationships between the triangles.

Tip

Create the shape with tangram pieces, then draw your outline. You can tape pieces together before you trace around them.

Build Your Understanding

Create a Tangram Puzzle

You Will Need
- complete set of tangram puzzle pieces

As you've learned in the last lesson and this lesson, tangrams can be used to create a great variety of shapes, such as polygons, letters, and pictures.

1. Create a tangram puzzle for classmates to solve. Follow these rules when creating your puzzle:
   a) You must use all seven tangram pieces.
   b) Pieces cannot be stacked or overlapped.
   c) Each piece must touch at least one other piece.

2. a) Draw an outline of your puzzle on a piece of paper.
   b) On a separate sheet of paper, draw the solution to the puzzle.

3. Exchange your tangram puzzle with a classmate or classmates.

Lesson 8: Exploring Tangrams Further

What Did You Learn?

1. Talk to the person who solved your puzzle. How difficult did he or she find it?

2. What made it difficult or easy?

3. Which strategy or strategies did you find most useful for solving challenging tangram puzzles?

Practice

Math Problems to Solve

1. **a)** Use two tangram pieces to make a parallelogram.

   b) Record your solution.

2. **a)** Use three pieces to make another parallelogram.

   b) Record your solution.

3. Use the tangram pieces to make one alphabet letter.

4. Which of the following polygons can you make using all seven tangram pieces? Record your solution(s).

   a) a triangle

   b) a rectangle that is not a square

   c) a parallelogram that is not a square

Extension

5. Use the tangram pieces to make each letter of your first name. Draw each letter of your name on a seperate piece of paper, and share it with the class.

Technology

Using a paint application, create and repeat regular polygons to tile a plane. Do Internet research on tessellations involving regular polygons. Experiment with regular polygons to see which ones tessellate and which do not. Use the text tool to explain your findings.

DATA MANAGEMENT AND PROBABILITY

Lesson 9
Probability

JOURNEY NOTE

PLAN:
You will learn the meaning of the following probability words: best/worst; probable/improbable; more likely/equally likely/less likely; always/never. You will use these words to complete sentences that describe situations in the board game Snakes and Ladders.

DESCRIPTION:
Have you ever played the board game Snakes and Ladders? You probably didn't know that Snakes and Ladders came to us from a game called Mosksh-Patamu, which was invented in ancient India.

The Taj Mahal in Agra, India

Get Started

In Snakes and Ladders, players take turns rolling a number cube. A player moves his or her counter around the board the number of spaces shown on the number cube. If the player lands at the base of a ladder, he or she can move up the ladder to the space at the top of the ladder. If the player lands on the head of a snake, he or she slides down the board to the space at the snake's tail. To win the game, a player must land exactly on 100.

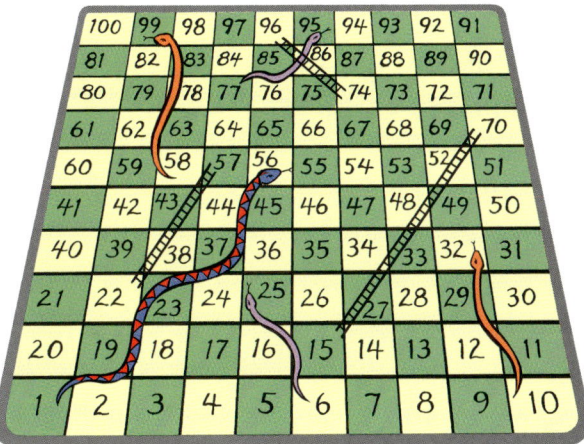

1. **a)** With a partner, examine the game board closely.

   b) Which are the good board spaces to land on?

   c) Which are bad ones? Why?

Lesson 9: Probability

2. Discuss answers to the following questions with your partner:
    a) Which ladder allows a player to move forward the farthest?
    b) Which snake is the most dangerous? Why?
    c) How many snakes are there? How many ladders are there?
    d) Is a player more likely to land on the head of a snake or on the base of a ladder? Explain your reasoning.
3. Share, compare, and discuss your answers with another pair of students.

Build Your Understanding

Understand Probability Words

You Will Need
- set of number cubes
- dictionary
- game board of Snakes and Ladders (optional)

1. When discussing situations that can happen in a board game, you might use words such as
    - best/worst
    - probable/improbable
    - more likely/equally likely/less likely
    - always/never

   In a small group, discuss what the words above mean. Show that you know the meaning of each word or phrase by using it in a sentence. For example, "It is improbable that it will snow today because there is not a cloud in the sky and it is the middle of June."

Journal
To make sure you understand the probability words, write definitions in your own words. Give an example sentence with each definition.

Tip
If you're unsure of a word's meaning, look it up in a dictionary.

Chapter 6: Math in Monuments and Marvels

2. Working with your group, select the word from the bulleted list in question 1 that best completes each Snakes and Ladders sentence below. A number of words or phrases may correctly complete some sentences. Be prepared to defend your word choices.

   a) Landing exactly on 100 is _____.

   b) Rolling four 6s in a row is _____.

   c) Rolling a 3 or a 5 is _____.

   d) Landing on the base of a ladder _____ means the player moves up the ladder.

   e) Landing on only even numbers is _____.

   f) Landing on the tail of a snake _____ allows the player to move up the snake.

   g) The chance a player will land on the head of at least one snake during the game is _____.

   h) If a player lands on more snakeheads than bases of ladders, his or her chances of winning are _____.

   i) That a player will land on the base of a ladder more often than on the head of a snake in the first move is _____.

Journal
Revisit the definitions you wrote earlier. Revise them based on new ideas you learned from the activity. You may wish to write example sentences from Snakes and Ladders game situations.

Vocabulary
improbable: Not likely to happen
probability: A number or statement that tells how likely it is that an event will take place
probable: Likely to happen

What Did You Learn?

Meet with another group to share, compare, and discuss your completed sentences.

1. Take turns explaining why you chose the word or phrase you did for each sentence.

2. Is it possible to use several different words or phrases to complete some sentences? Which sentences? Why?

Practice

Extension

1. Make up a sentence using each of the following words or phrases correctly. Use situations from your own life to show that you know what the words mean.

   - best
   - probable
   - more likely
   - always
   - worst
   - improbable
   - equally likely
   - never
   - less likely

2. **a)** Read your sentences from question 1 to a partner, but leave out the probability word or phrase from each sentence.

   b) Have your partner complete each sentence.

   c) Compare your answers to see if they are the same.

   d) If they are different, each of you must defend your word or phrase choice.

3. Make up Snakes and Ladders sentences, like the ones in Build Your Understanding, for classmates to complete. Base the sentences on your experiences playing the game. Make sure you know how to correctly complete each sentence.

4. **a)** Look at weather forecasts in your local newspaper.

   b) Cut out forecasts in which the writer uses words or phrases from question 1.

   c) Present these examples to the class.

   d) Explain what each sentence means and why the writer used these particular words.

Technology

Use a desktop publishing program to create a mind map related to probability. Include probability words as well as activities from everyday that involve chance and/or probability.

Chapter 6: Math in Monuments and Marvels

DATA MANAGEMENT AND PROBABILITY
NUMBER SENSE AND NUMERATION

Lesson 10

Exploring Probability Further

JOURNEY NOTE

PLAN:
You will play a game called the Odd-Even Game. You will learn about probability and use this knowledge to develop a strategy to improve your game performance.

DESCRIPTION:
Archeologists have found number cube games in the tombs of ancient Sumeria in Egypt. Most number cubes were made of bone or ivory. Number cubes found near Rome, and thought to be made about 900 B.C., are similar to the number cubes of today, with the opposite faces adding up to 7.

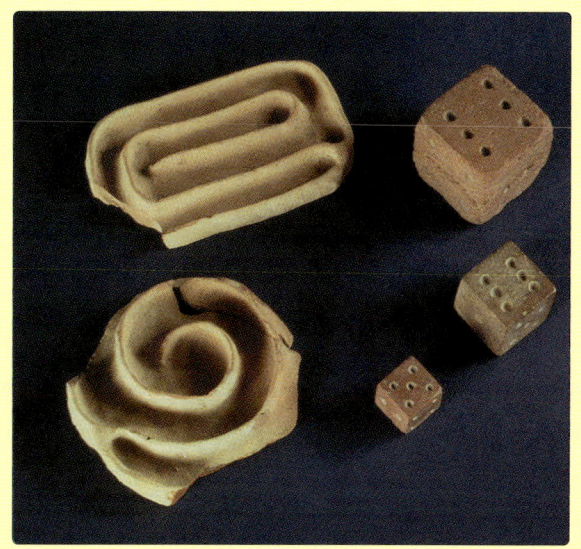

Ancient number cubes

Get Started

A Math Game to Play

You are going to play a numbers game called the Odd-Even Game.

What You Need
- one odd-even chart for each player
- 22 counters per player
- one pair of six-sided number cubes for each pair of students

Play the game with a classmate.

Lesson 10: Exploring Probability Further 295

How to Play the Odd-Even Game

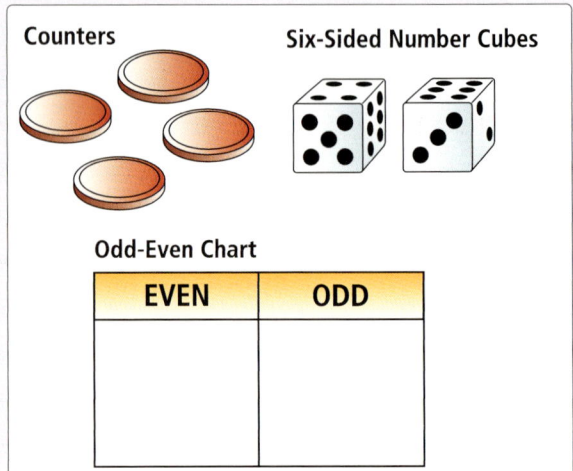

1. Divide your 22 counters into two categories on your chart, one called "Odd" and the other "Even."
2. Take turns rolling the number cubes.
3. After a player roles the number cubes, he or she calculates the product of the values on the faces of the cubes.
4. If the product is an even number, the player removes a counter from his or her "Even" column.

   If the product is an odd number, the player removes a counter from his or her "Odd" column.
5. If you roll a product for a category in which all counters have been cleared, you lose a turn.
6. The first player to remove all counters for both categories wins.

Vocabulary

even number: A whole number with 2 as a factor. The number can be divided by 2.
odd number: A number that does not have 2 as a factor. The number cannot be divided by 2.
product: The answer in a multiplication calculation

Journal

Make note of any patterns you notice in the number of even-number products you obtain versus the number of odd-number products. How could you use this knowledge to improve your strategy for playing the game?

Build Your Understanding

Improve Your Game Strategy

You Will Need
- one odd-even chart for each player
- 22 counters per player
- set of six-sided number cubes
- Product Outcomes table

Chapter 6: Math in Monuments and Marvels

1. Discuss with your partner how you might divide your 22 counters so that you stand a better chance of winning the game.

2. Divide your counters according to the new strategy. Play the game from Get Started again.

3. Play the game a third time. However, for each turn, record the number on each cube and the product of the two numbers in the correct column of the Product Outcomes table.

Even Products	Odd Products

Product Outcomes					
Even Products			Odd Products		
Number on Cube 1	Number on Cube 2	Product	Number on Cube 1	Number on Cube 2	Product

What Did You Learn?

Meet to share and discuss your game experiences, game strategies, and the results on your Product Outcomes table.

1. How many times did you get an even-number product?

2. How many times did you get an odd-number product?

3. How could you use what you know to develop a better strategy for organizing your piles of odd and even counters the next time you play?

Lesson 10: Exploring Probability Further

Practice

Extension

1. Try the strategy you identified in question 3 of What Did You Learn? and comment on how well it works.

2. Working with your partner, make a list of all possible combinations of rolls and the product of each roll. Organize the possibilities into a chart with Odd and Even columns.

Odd Products	Even Products
1 x 1 = 1	1 x 2 = 2
1 x 3 = 3	2 x 2 = 4

3. a) From your chart, what can you conclude about multiplying even and odd numbers together?

   b) What can you conclude about multiplying even numbers with even numbers?

   c) What can you conclude about multiplying odd numbers with odd numbers?

Technology

Brainstorm a list of games you know or have played. Use a draw application to create a Venn diagram identifying Games of Skill and Games of Chance. Place the games appropriately in the diagram, including in the overlapping area. Use the text tool to explain the reason for your placements.

Show What You Know

Review: Lessons 9 and 10, Probability

1. Create a simple game of probability.
2. Write the rules.
3. Play the game with a partner.

Chapter 6 Chapter Review

1. Choose two rooms in your school. Using what you have learned in this chapter, estimate their heights. Explain how you got your estimates.

2. a) A triangle has ■ sides.
   b) A triangular prism has ■ edges.
   c) A rectangle has ■ sides.
   d) A rectangular prism has ■ edges.
   e) If a pentagon has ■ sides, how many edges does a pentagonal prism have?

3. This is a pentagonal prism.
   a) How many edges are there?
   b) How many vertices are there?

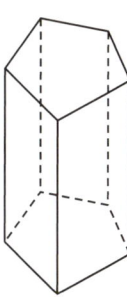

4. Which net(s) can be folded to make a rectangular prism?

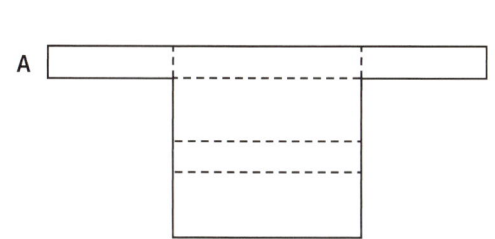

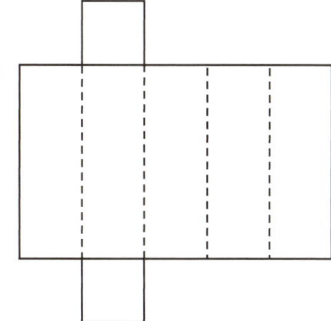

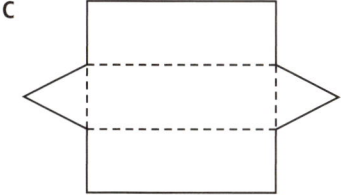

5. Name the solid that each net could make.

A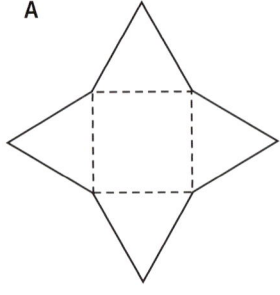

B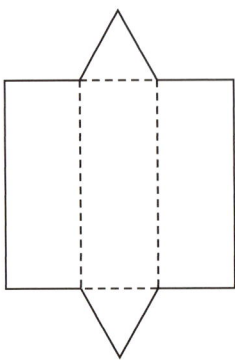

C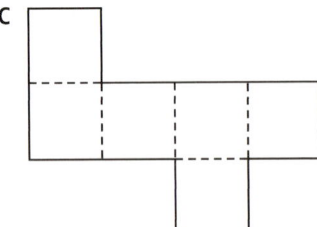

D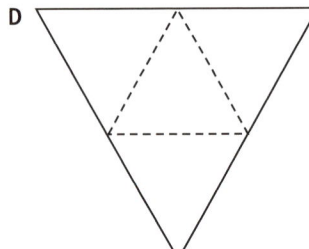

6. In what way(s) are pyramids and prisms similar?

7. Label all the parts of this three-dimensional figure.

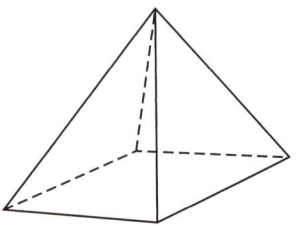

8. Using the faces below, draw three-dimensional figures and name them.

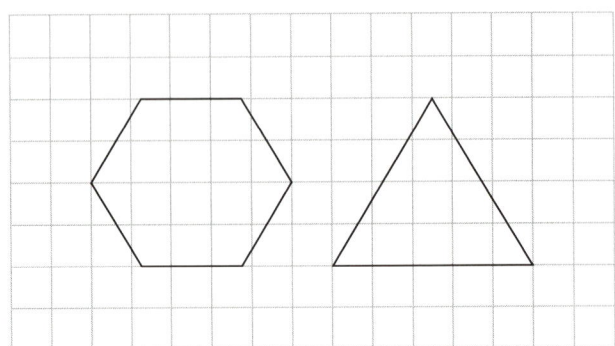

9. Determine the number of cubes in each shape below. Explain how you got your answers.

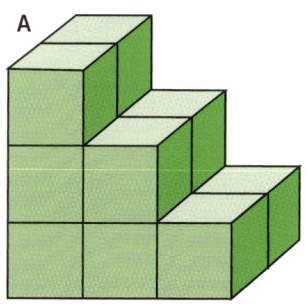 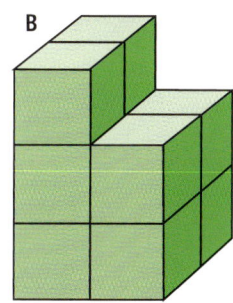

10. Use a complete set of tangram puzzle pieces to make an animal or object.

11. Trace these two triangles and then cut them out. Make three different shapes by combining the two triangles. Name the different polygons you created.

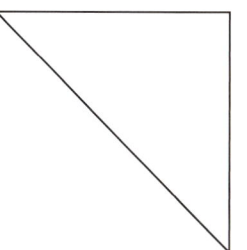 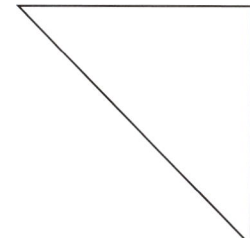

12. From the list below, use the most appropriate word to complete the statements.
- probable
- improbable
- always
- never

a) If there are large storm clouds in the sky, it is _____ that it will rain.

b) If the car is out of gas, it is _____ that it will start.

c) When you have the flu, you _____ want to play tag.

d) If you add 2 plus 2, you _____ get 4.

Chapter 6

Chapter Wrap-Up

You have learned about the great monuments of early civilizations. Many of these important structures come in interesting three-dimensional shapes and are decorated by attractive two-dimensional designs and patterns. Obelisks, for example, were dedicated by ancient Egyptians to their sun god and placed in pairs at the side of temple entrances. What pyramid or prism shapes do you see in the obelisk?

It's time to apply the knowledge and skills you've learned in this chapter to complete this project. On your own, design, construct, and decorate a monument to celebrate an important person from an early civilization.

1. The monument must be a pyramid or prism.
2. Construct the monument out of paper.
3. Decorate the monument with different polygons.

Provide this written information with your monument:
- the name of the type of pyramid or prism you have used and the name of the polygons that form its faces
- all dimensions of the monument, including its height and the area of the base on which it stands
- the names of the polygons you used to decorate the monument
- a net of the pyramid or prism
- a three-dimensional drawing of the monument

Obelisk at Karnak, Egypt

NUMBER SENSE AND NUMERATION

Problems to Solve

Here are some more fun problems for you to solve. You will be given a helpful problem-solving strategy for the first two problems. For the last three, you get to choose a strategy to use.

Problem 10

Making Equations

STRATEGY: WRITE AN EQUATION
Sometimes, coming up with a math equation can help you solve a problem.

OBJECTIVE:
Demonstrate an understanding of multiplication and addition

Problem

Imagine 9 frogs live in a pond. Each frog eats 25 insects a day. How many insects in total do all the frogs eat in a week? Use the "Write an Equation" strategy to solve this problem.

Tip

Problem-Solving Steps
1. Understand the problem.
2. Pick a strategy ("Write an Equation").
3. Solve the problem.
4. Share and reflect.

Problems to Solve

Reflection

1. What information did you know about the problem before you solved it?
2. What did you need to figure out?
3. What type of calculations did you need to do? How did you create an equation?
4. Do you think the "Write an Equation" strategy was a good strategy to use? Why or why not?
5. What other strategy might you use to solve this problem? Explain why.

Extension

1. How many insects do all the frogs eat in 4 weeks? Show your work using pictures, numbers, and words.
2. Make up a similar question about another kind of pond animal. Give your problem to a classmate to solve.

Journal

Record what you know about writing equations.

NUMBER SENSE AND NUMERATION

Problem 11
Using Logic

STRATEGY: USE LOGICAL REASONING

Logical reasoning helps you think about and solve a problem using information that you know.

OBJECTIVE:

Explain your thinking when solving problems

Problem

You Will Need
- four different colours of blocks

Tomas, Noma, Jessica, and Anil take a trip to the amusement park. The 4 friends decide to go on the log ride. Each log has 4 seats. Tomas doesn't want to sit in front. Noma sits ahead of Jessica. Jessica sits behind Anil. How might the friends be sitting in the log? Use the "Use Logical Reasoning" strategy to solve this problem.

Reflection

1. What information did you know about the problem before you solved it?
2. What did you need to figure out?
3. How could you use the blocks to solve the problem?
4. Which position did you start with? Why?
5. Look back at your answer. Why does it make sense?
6. Do you think the "Use Logical Reasoning" strategy was a good strategy? Why or why not?

Problems to Solve

Extension

1. Look at how your class is arranged. Choose one part of your classroom, maybe a row or a group of students. Make up a problem about how this part of your class is arranged.

2. Make up a problem about where your family members sit at dinner time.

3. Give your problems from questions 1 and 2 to a classmate to solve.

Problem 12
Estimating Lengths of Bathroom Tissue

STRATEGY: YOUR CHOICE

OBJECTIVE: Estimate lengths

Problem

You Will Need
- rolls of bathroom tissue

Have you ever looked closely at a roll of bathroom tissue? Here are two problems for you to solve that have to do with bathroom tissue:

306 Chapter 6: Math in Monuments and Marvels

1. Estimate the number of squares of bathroom tissue needed to make a length of 1 m, 10 m, and 100 m.

2. How many metres do you think 1 roll of bathroom tissue would extend? Explain how you came up with your estimate.

Reflection

1. What information did you know about the problem before you solved it?

2. What did you need to figure out?

3. Do you think the strategy you chose was a good strategy to use? Why or why not?

4. Why do you think your estimates make sense?

Extension

Look around your classroom. What could you make estimation problems about? Create a problem for a classmate to solve.

Problem 13
A Division Problem

STRATEGY: YOUR CHOICE

OBJECTIVE:
Use division skills

Problem

Sasha is taking a car trip across the province with her family. They need to travel 893 km. The average speed limit is 80 km/h. How long do you think it will take to travel the entire distance? Remember that they will need to stop for lunch and for washroom breaks. Use a strategy of your choice to solve this problem.

Reflection

1. What did you need to figure out to solve this problem?
2. List the steps that you needed to follow to solve this problem.
3. Which strategy did you pick? Why?
4. Do you think your strategy was a good strategy to use? Why or why not?
5. Share your results with another classmate. Are your results the same or different? Why might your answers be different?

Extension

1. How far would Sasha's family travel if they were driving 90 km/h for 12 h but stopped for a 45-min lunch break?
2. How far would they travel if they didn't stop for lunch?
3. What is the difference between the two distances? Show your work.

NUMBER SENSE AND NUMERATION

Problem 14

Working With Whole Numbers

STRATEGY: YOUR CHOICE

OBJECTIVE:
Solve two-step problems involving whole numbers

Problem

Imagine that the students in your school are going to take the subway to the zoo. There are 400 students in your school. The subway has 60 rows of seats in each car. What would be the fewest number of subway cars you would all need if no more than 3 students sat in a row? Don't forget to give the teachers a place to sit. There will be 1 teacher for every 10 students. Use a strategy of your choice to solve this problem.

Reflection

1. What information did you know about the problem before you solved it?
2. What did you need to figure out?
3. How many steps were needed to solve this problem?
4. Which strategy did you use?
5. Do you think this strategy was a good choice for this problem? Why or why not?
6. Compare your answer to the problem with another classmate's.

Extension

Use a calculator to check your work.

Problems to Solve

Unit 3
Math in Energy

The citizens of Mathford are concerned about how much energy they use every day and how much energy their town has available. It's important for them to know this information because energy is a valuable resource and supplies might run out if it is used wastefully. In Unit 3, you will help the people of Mathford measure their energy use. You will also use your math skills to find ways to conserve energy.

In Chapter 7, you will look at how math is used to measure electricity. You will learn how to calculate electricity in watts, and how to read an electricity meter. In the Chapter Wrap-Up, you will analyze data and create math questions about the data.

In Chapter 8, you will learn how people have stopped using wood to heat their homes and now use other sources of heat in order to conserve trees. In the Chapter Wrap-Up, you will analyze graphing data.

In Chapter 9, you will discover the importance of saving energy in your community and learn about some of the ways you can conserve energy. In the Chapter Wrap-Up, you will use what you have learned to prepare a report about energy conservation.

Chapter 7

Electric Math

The people of Mathford need to use math to read their electricity bills. In this chapter, you will learn how to measure electricity so you will know how much energy you are using every day.

In this chapter, you will
- collect data on electricity use and convert data to fractions
- calculate the mode and mean of data
- create and use a formula to calculate data
- calculate the cost of energy, and analyze and compare energy use
- add and subtract amounts of money
- learn how to read meter dials and multiply by 10, 100, and 1000
- find the perimeter and area of irregular polygons

At the end of this chapter, you will analyze data about electricity use and create math questions about the data.

1. What happens to the number 3 when it is multiplied by 10, 100, and 1000? What rule can you state?
2. Work with a classmate, and review what a fraction is.
3. How can you find the perimeter and area of a shape?

Chapter 7: Electric Math 311

DATA MANAGEMENT AND PROBABILITY

Lesson 1

Analyzing Energy Data

ENERGY NOTE

PLAN:
You will collect data on how you use electricity, convert the data to fractions, and calculate the mode and mean of your data.

DESCRIPTION:
Can you see energy? Can you feel it, touch it, or smell it? The people of Mathford want to know when they are using energy, so they need the answers to these questions.

Get Started

1. Look at these pictures.
   a) Where is electrical energy being used?
   b) How do you know?
   c) What activities use more energy than others?
   d) How could you find out?
   e) Discuss your answers with a partner.

Vocabulary

mean: The average of a set of numbers. To calculate the mean, add up all the numbers and divide by the number of choices given. For example, if you had 1 item using 3 energy forms, 1 item using 2 energy forms, and 1 item using 1 energy form, the mean would be 2 energy forms. 3 + 2 + 1 = 6   6 ÷ 3 = 2

mode: The most frequent number in a group. If you have 5 items using heat energy, 5 items using sound energy, 3 items using motion energy, and 1 item using light energy, the mode is 5.

Chapter 7: Electric Math

The following chart shows how many small appliances three families use.

Family	Toaster	Television	Electric Razor	Radio	Iron	Hair Dryer	Electric Toothbrush	Total
A	2	1	1	3	1	4	2	
B	4	0	2	2	4	3	4	
C	1	2	0	1	1	2	0	
Total								

2. Copy and complete the chart in your notebook.
3. Find and record the mode of the data for each appliance.
4. Find and record the mean of the data for each family.
5. What fraction of the total number of each small appliance does each family use?

Tip

A fraction is a part of a whole number. The top number is the numerator, which is the number of parts being referred to. The bottom number is the denominator, which is the total number of parts.

Build Your Understanding

Count Energy Forms

Electrical energy takes different forms. Energy forms are heat, light, motion, and sound. For example, a light bulb transforms electricity into both heat and light.

1. Look at the table on the next page. Discuss with a partner which energy forms are related to each item.
2. Can one item transform electricity into more than one energy form? Explain your answer.
3. Copy the chart into your notebook, and check off the energy form(s) related to each item.
4. Add up the total in each column. Order the totals from greatest to least.
5. Calculate the mode and mean of your totals.
6. What fraction of the items generates heat? light? motion? sound?

Lesson 1: Analyzing Energy Data

Electrical Item	Energy Form			
	Heat	Light	Motion	Sound
doorbell				
television				
toaster				
electric razor				
radio				
light bulb				
Total				

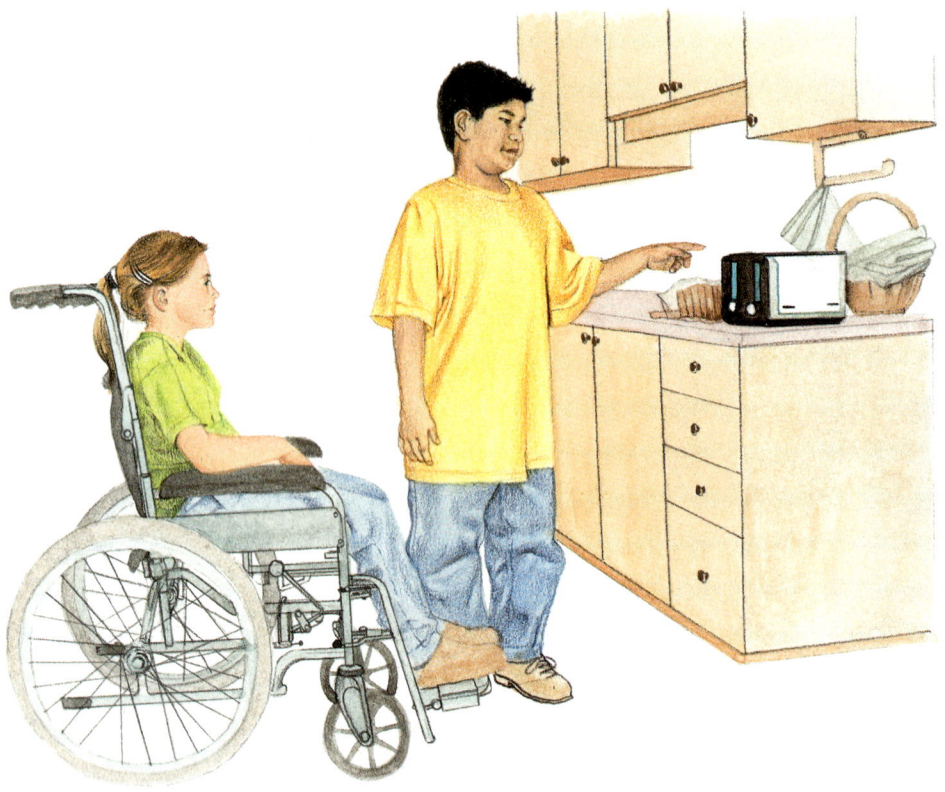

What Did You Learn?

1. Is one energy form used more often than another? Why might this be? Share your thoughts with another pair of students.

2. How did the table help you organize and display the data? Explain your answer.

3. How did you find the mode? How did you find the mean?

Chapter 7: Electric Math

Practice

Order these numbers from least to greatest:

1. 79 564, 617 578, 674 629, 98 543, 672 857
2. 349 564, 507 280, 341 665, 345 781, 98 407
3. 128 827, 205 650, 128 459, 250 543, 123 995

Calculate the mode and the mean for each set of numbers:

4. 36, 47, 82, 47, 36, 93, 36, 21, 17, 7
5. 4, 103, 871, 3, 103, 24, 100, 24, 103
6. 634, 1, 323, 634, 34, 207, 206, 634, 2

Write each equation in fractions, without words, and find the missing terms.

7. 6 tenths + 3 tenths = ■ tenths
8. 4 hundredths + 4 hundredths = ■ hundredths
9. 5 tenths + 2 hundredths = ■ hundredths

Vocabulary

watt: A unit used to measure the rate of electricity use. The metric symbol for watt is W. The lower the watts, the less electricity used. The amount of energy used is measured in watts per hour (watt hours or Wh) or kilowatts per hour (kilowatt hours or kWh). 1 kW = 1000 W

wattage: The amount of electricity

Extension

10. **a)** Suggest and add other electrical items to the table in this lesson's Build Your Understanding.

    b) Now calculate a new mode and mean for heat, light, motion, and sound.

11. **a)** Light bulbs produce both light and heat. Record how many light bulbs you use in your home. Make a chart like the one to the right. Record the wattage for each light bulb. You can look at the photo on page 316 to help you. An example has been done for you. What do you notice about your data?

    b) Meet as a class and share your data. Use your data to create a class comparison chart.

Wattage	Number of Light Bulbs
40 W	5

Technology

Use a database computer program to create a database of objects that use energy. Then sort the data, analyze the data, and draw conclusions.

Lesson 1: Analyzing Energy Data

Lesson 2

Working With Data and Formulas

ENERGY NOTE

PLAN:
You will gather data and then use a formula to multiply the data.

DESCRIPTION:
How much energy does a light bulb use? How can the people of Mathford find out, without turning off the lights? Did you know that lights consume about 20 percent of all energy used in Canada?

Get Started

Look at the photographs of these light bulbs. Each bulb uses a certain number of watts.

It is a school night in Mathford, and many Grade 5 students are doing their homework.

- Mandar, Jamar, and Shefali are each using desk lamps. Each desk lamp has one 25 W light bulb.
- Tavia and Anil are each sitting under overhead lights. Each overhead light has one 40 W light bulb.
- Dan, Sasha, Tomas, and Noma are each sitting beside table lamps. Each table lamp has one 60 W light bulb.

316 Chapter 7: Electric Math

1. What is the total wattage of the bulbs being used by these students?

2. What fraction of the total watts is being used by Mandar, Jamar, and Shefali?

3. Now write the fraction of watts being used by Tavia and Anil.

4. Now write the fraction of watts being used by Dan, Sasha, Tomas, and Noma.

Build Your Understanding

Calculate Watts

1. Imagine that the students finished doing their homework but then forgot to turn off the lights when they went to bed. Instead, they turned them off in the morning when they woke up, which means that the lights were on all night!
   - Mandar, Jamar, and Shefali each did homework for 3 h and slept for 9 h.
   - Tavia and Anil each did homework for 2 h and slept for 8 h.
   - Dan, Sasha, Tomas, and Noma each did homework for 3 h and slept for 8 h.

   In your notebook, copy and complete the chart below to find the total number of watts used. Use the information from Get Started.

	Number of Bulbs	Watts per Bulb	Total Watts	Total Hours Used in One Day	Total Watt Hours Used
Mandar, Jamar, Shefali					
Tavia, Anil					
Dan, Sasha, Tomas, Noma					

2. How many total watt hours did the students use?

3. How many watt hours would be used if all of the students did this for 3 nights?

Lesson 2: Working With Data and Formulas

4. How many watt hours would the students have used if they had turned off the lights after they had done their homework? How many watt hours would they have saved? Show your work.

What Did You Learn?

1. What is a formula and how does it help you with calculations?

2. What strategies did you use to calculate the total number of watt hours used by these students? Share and compare your strategies with a classmate. What similarities and differences do you notice? Try to determine a formula that would make these calculations easier.

3. Explain the formula that you came up with, in pictures, numbers, and words.

Practice

1. Keep track of how many hours a day the lights are on in your home. Count how many watt hours are used to light your home for a day. Share your findings with the class.

2. If you switched one of the 60 W light bulbs in your home to a 25 W light bulb, how much electricity would you save? How much would you save if you switched 5 of them? Is this a practical way to save electricity? Why or why not?

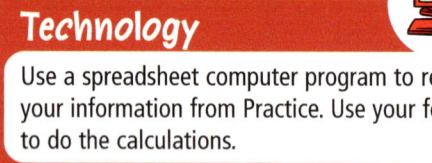

Technology
Use a spreadsheet computer program to record your information from Practice. Use your formula to do the calculations.

DATA MANAGEMENT AND PROBABILITY
NUMBER SENSE AND NUMERATION

Lesson 3

Examining and Organizing Energy Data

ENERGY NOTE

PLAN:
You will gather data and then use your formula from Lesson 2 to multiply the data.

DESCRIPTION:
Many devices around the home use energy. The people of Mathford like their appliances and use them to do helpful tasks around the house. How much electricity does a household appliance use?

Get Started

1. Why is it important to know how much electricity you use?

2. With a partner, choose one item from the chart on the next page. Estimate how many hours your family uses this item in a day, a week, and a month.

3. Now calculate how much electricity that item uses in a day, a week, and a month. Share your results with the class.

Tip

You can use the same method that you used in Lesson 2 to calculate the data. Use a calculator to check your answers.

Lesson 3: Examining and Organizing Energy Data

Television	200 W	Electric Shaver	15 W	Kettle	1500 W
Dishwasher	1300 W	Toaster	1150 W	Food Processor	390 W
Iron	1000 W	Clothes Dryer	5000 W	Compact Stereo	18 W
Computer	200 W	Curling Iron	9 W	Refrigerator	500 W
Microwave Oven	1000 W	Electric Frying Pan	1150 W	Can Opener	90 W
Sewing Machine	75 W	Power Saw	275 W	VCR	40 W
Hair Dryer	1000 W	Electric Toothbrush	10 W	Coffee Maker	900 W
Stove	12 500 W	Blender	100 W	Clock Radio	8 W
Washing Machine	500 W	Vacuum Cleaner	800 W	DVD Player	50 W

Build Your Understanding

Calculate Energy

1. Use the data in the chart from Get Started. Choose six items from different areas of your home. Write down the six items and their wattage in an energy chart like the one below. Estimate and record in the proper column how many hours each item is used.

Item	Wattage	Estimate of Total Hours Used in a Day	Number of Watt Hours per Day	Number of Watt Hours per Week	Number of Watt Hours per Month	Number of Watt Hours per Year
Totals						

2. Complete the rest of the chart. You may use a calculator.

3. Compare your chart with that of a classmate. How are they similar and how are they different?

4. Create a graph showing the energy use of the six items you chose from the chart. Write some questions that can be answered using the data in your graph.

5. Compare your graph with your partner's graph. What questions can you ask your partner about the information in his or her graph?

Technology

Use a word processing program to create some word problems using the data from Get Started. Print out your word problems, and create answers for each on a separate piece of paper. Challenge your classmates to solve your problems.

What Did You Learn?

1. What did your energy chart show you about electricity use?

2. How much electricity could you save if you used each item on your chart 3 h less per year? Show your work.

3. How did your line graph help you organize your data?

Journal

Describe the process you used to calculate how much electricity you were using.

Lesson 3: Examining and Organizing Energy Data

Practice

1. How might you save on electricity in your household? What recommendations would you make to your family?

2. Using your chart from this lesson's Build Your Understanding, calculate your family's mean wattage per day. Explain how you got your answer.

Show What You Know
Review: Lessons 1 to 3, Analyzing Data

1. Explain how to calculate the mode and the mean for a given set of numbers. Use examples to support your answer.

2. Copy a chart like this one in your notebook:

Hours Used	25 W Light Bulb	40 W Light Bulb	60 W Light Bulb
9 h			
7 h			
5 h			

a) Calculate the watt hours used for each light bulb.

b) What patterns do you notice in your chart?

Chapter 7: Electric Math

DATA MANAGEMENT AND PROBABILITY
NUMBER SENSE AND NUMERATION

Lesson 4
Calculating Energy Cost

ENERGY NOTE

PLAN:
You will calculate the energy cost of household appliances.

DESCRIPTION:
Every appliance has two prices. The first price is the purchase price. The other price is how much energy the appliance uses, which is called an energy cost or energy rating. You need to do some math to find this cost.

Get Started

You Will Need
- calculator

Labels can be very useful. Appliance manufacturers use "EnerGuide" labels on appliances to tell us how much energy an appliance uses. Why is knowing how much electricity an appliance uses important to consider when you are making a purchase? Share your ideas with the class.

Tip
The lower the EnerGuide rating, the less electricity the appliance uses.

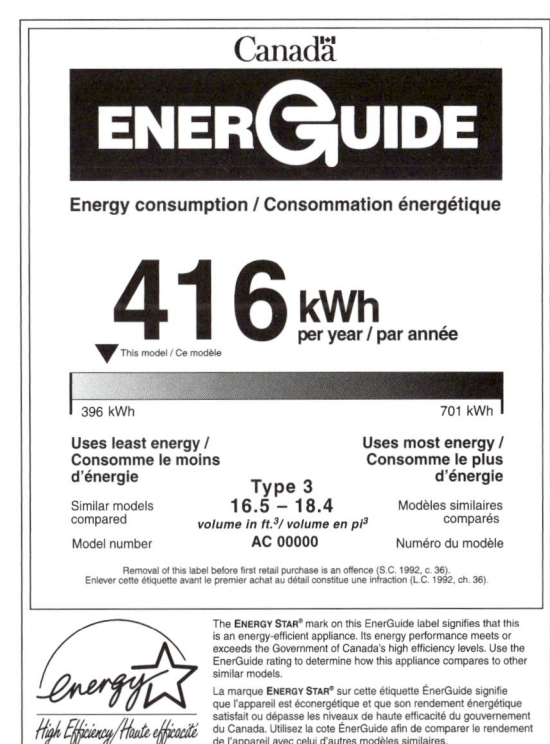

Lesson 4: Calculating Energy Cost   323

Look at the chart below.

Major Electrical Appliances	Lifetime Expectancies in Years
dishwasher	6
refrigerator	8
range (stove)	9
washing machine	8
clothes dryer	7
freezer	9

1. Imagine 1 kWh of electricity costs 8¢. With a partner, look at the chart below.

Appliance	Model	EnerGuide Rating (kWh)	Price of 1 kWh of Electricity	Appliance Life (Years)	Annual Energy Cost ($)	Total Cost of Energy Over the Life of the Appliance
dishwasher	model 1	483	8¢			
	model 2	426	8¢			
refrigerator	model 1	414	8¢			
	model 2	345	8¢			
range (stove)	model 1	396	8¢			
	model 2	379	8¢			
washing machine	model 1	370	8¢			
	model 2	259	8¢			
clothes dryer	model 1	430	8¢			
	model 2	394	8¢			
freezer	model 1	448	8¢			
	model 2	416	8¢			

2. Copy and complete the data for one appliance from the above chart. Calculate the annual energy cost of the appliance. Then calculate the cost of energy over the life of the appliance. You may use a calculator and the Tip box on the next page to help you.

Chapter 7: Electric Math

Tip

You can find out the amount of energy an appliance uses in a year from its EnerGuide rating. Multiply this by the expected life of the appliance to give you the total kilowatts it will use during its lifetime. Then multiply the number of kilowatts by the price for one kilowatt. Remember to divide your answer by 100 to convert the total cost from cents to dollars.

kilowatts used in a year x number of years
x cost per kWh (in cents) ÷ 100
= energy cost (in dollars)

Vocabulary

kilowatt hour: A unit for measuring electricity equal to 1000 Watts of electricity used in one hour. The metric symbol for kilowatt hour is kWh.

Build Your Understanding

Calculate the Cost of Energy

1. Use the charts on the previous page to calculate the energy cost of each appliance on the chart.

2. For each appliance, which model has the lowest total cost of energy over the life of the appliance?

3. Find the mean total energy cost for each appliance. Refer to the glossary to review the definition of mean.

Journal

Use pictures, numbers, and words to explain how to calculate the energy cost of an appliance.

What Did You Learn?

1. What observations can you make by looking at your completed chart?

2. Use pictures, numbers, and words to explain how you calculated the mean total energy cost for each appliance.

Lesson 4: Calculating Energy Cost

Practice

You Will Need
- appliance catalogues and flyers

1. Look at the prices of appliances in catalogues and flyers. Calculate the mean of 5 refrigerators and 3 clothes dryers. Don't forget to calculate and add the taxes when determining the total price of an item.

2. Choose one item from each appliance category in Get Started to buy from the catalogue.
   a) Use the EnerGuide rating given in the chart to calculate the energy cost of these items. (You can choose the EnerGuide rating of Model 1 or 2.)
   b) Now calculate the total cost of each appliance, including the catalogue retail price (with taxes), as well as the energy cost.
   c) Calculate the difference between the catalogue retail price and the energy cost of each item.

3. Choose two different microwaves from a catalogue or flyer. Label them Model 1 and Model 2. Assume that the more expensive microwave has an EnerGuide rating of 400 kWh and the other microwave has an EnerGuide rating of 380 kWh. Both microwaves have a life expectancy of 8 years.
   a) Calculate the energy cost of these items.
   b) Now calculate the total cost of each microwave, including the catalogue retail price (with taxes), as well as the energy cost.
   c) Which model would you recommend to your family? Why?

4. Calculate the mean and then the mode in each set of numbers:
   a) 62, 371, 52, 62, 8, 103, 62, 5, 62
   b) 3, 6, 12, 888, 12, 14, 2, 1, 9
   c) 56, 314, 987, 519, 905, 56, 7, 56

Technology

Create an advertising brochure using a desktop publishing program. Advertise each of the appliances that you feel are the best buys (from the data in this lesson), and include important information that a customer would want to know.

NUMBER SENSE AND NUMERATION

Lesson 5

Analyzing an Energy Graph

ENERGY NOTE

PLAN:
You will analyze a graph about energy use and conduct a survey.

DESCRIPTION:
Some appliances use less electricity than other appliances. The citizens of Mathford know this, and they are eager to find out which ones use less energy.

Get Started

Our homes are filled with appliances that help make our lives easier. They all use different amounts of energy.

1. a) In groups of three or four students, brainstorm a list of appliances used in your homes.

   b) Survey each classmate in your group for the estimated hours per week that these appliances are used.

   c) Display your data on a chart like the one on the next page. The first section has been completed as an example.

Lesson 5: Analyzing an Energy Graph

Appliance	Student	Estimated Hours Used in One Week
microwave	Mandar	3
	Noma	2
	Tavia	0

d) Calculate the mean hours used for each appliance.

e) Now, using the means, rank the appliances in order from the one used most often to the one used least often.

Build Your Understanding

Analyze a Graph

With a partner, use the graph below to evaluate energy use of appliances in Canada for eight years.

1. Which appliance consistently uses the most energy?

2. Which appliance consistently uses the least energy?

Vocabulary

megawatt: A unit for measuring electricity equal to 1000 kW or one million watts. The metric symbol for megawatts is MW.

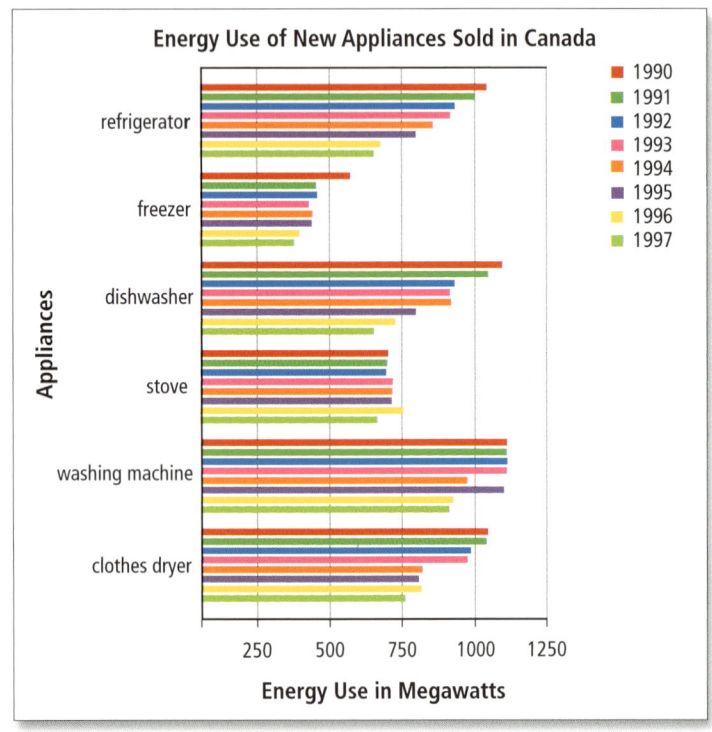

3. Which appliance has improved the most in becoming more energy efficient? How do you know?

4. Which appliance's use of energy has stayed about the same from 1990 to 1997? Why might that be?

What Did You Learn?

1. The graph shown in this lesson's Build Your Understanding is another type of bar graph. The bars are horizontal. They are in groups of eight to represent the eight years of data. How else could the data on this graph be presented to make the information meaningful to the reader? Would it be easier to understand? Why or why not?

2. What else can you learn about appliance use and energy use from this graph? Share your answers with a partner.

Practice

1. Rank the appliances from the graph in order of energy use.

2. Calculate the mean and the mode of the appliances, in terms of their energy use.

3. Survey 10 friends, neighbours, and family members on what appliance they think uses the most energy. Rank their responses.

4. How do the survey results compare to the data on the graph?

Technology

How is electricity generated where you live? Use a search engine to locate and access Web sites to find out where and how your electricity is made.

Lesson 5: Analyzing an Energy Graph

Lesson 6

Working With Money

ENERGY NOTE

PLAN:
You will add and subtract amounts of money.

DESCRIPTION:
There's a sale on small appliances at the Mathford Department Store, and many people have gone to check out the prices. Some people may even buy a new appliance for their home.

Get Started

When you go shopping, what do you need to consider? Imagine that you have $25.00 and you need to buy some school supplies. The prices, including taxes, are shown on the right.

1. With a partner, make a list of all the possible combinations of items you could purchase. Share your results with another pair of classmates.

2. Explain the problem-solving strategies you used to complete this task.

Chapter 7: Electric Math

Build Your Understanding

Make Change

A hotel has $100.00 to buy new small appliances. The chart lists the prices including taxes. Find as many combinations of items to purchase as possible. Create a chart to show all the possible combinations you find.

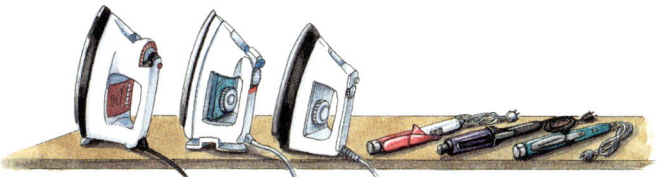

Clock Radio A	$ 58.00	Electric Shaver A	$ 35.00
Clock Radio B	$ 62.00	Electric Shaver B	$ 45.00
Clock Radio C	$ 49.00	Electric Shaver C	$ 27.00
Curling Iron A	$ 15.00	Iron A	$ 18.00
Curling Iron B	$ 25.00	Iron B	$ 25.00
Curling Iron C	$ 19.00	Iron C	$ 50.00

Remember to show your work so you can explain your choices.

What Did You Learn?

Check your answers with a partner. Recalculate if you disagree and can't come to a consensus. Then compare your answers with another pair of students.

Journal

What problem-solving strategies did you use? What was your largest combination? What was your smallest combination?

Practice

1. When you chose which items to buy, all you saw from the chart was the purchase price. What other information would you need to help you decide the total cost of each appliance, including the energy costs? How might that information affect your choices?

2. Does getting the most value for your money mean always paying the least amount of money? Why or why not?

Lesson 6: Working With Money

Show What You Know

Review: Lessons 4 to 6, Data Management and Money

1. Imagine that two vacuum cleaners are on sale for (1) $229.00 and (2) $320.00. The first vacuum cleaner uses 350 kWh of energy and the second one uses 300 kWh. Both vacuums have a life expectancy of 6 years. Which vacuum is the better buy? Use pictures, numbers, and words to show your work.

2. List the steps you followed to solve question 1. What did you find easy and what did you find most challenging? Why?

3. Share your answer and the steps you followed with a classmate.

4. Choose five appliances. Survey 20 friends, classmates, and family members to find out which appliance from your list they use the most in one week. Graph your data and share it with the class.

5. Write three questions that can be answered by looking at your graph. Provide answers to your questions.

6. Using an appliance catalogue or flyer, find and record different combinations of items that you could buy with $100.00. What strategies did you use to complete this task?

NUMBER SENSE AND NUMERATION

Lesson 7

Multiplying by 10, 100, and 1000

ENERGY NOTE

PLAN:
You will learn how to read meter dials and multiply by 10, 100, and 1000.

DESCRIPTION:
Electric meters keep track of how much electricity a household uses. Every home in Mathford has one of these meters, but not everyone in Mathford knows how to read it.

Get Started

You cannot see electricity, but you can measure it. Some homes have a precise instrument to read electricity. It's called the electric meter. It measures how many kilowatt hours a household uses.

An electric meter has four or five dials like the ones in the photo above. Some dials move clockwise while others rotate counterclockwise. The dials on your electric meter never stop moving, unless the power is cut off and you stop using all electricity.

Here's how to read an electric meter.

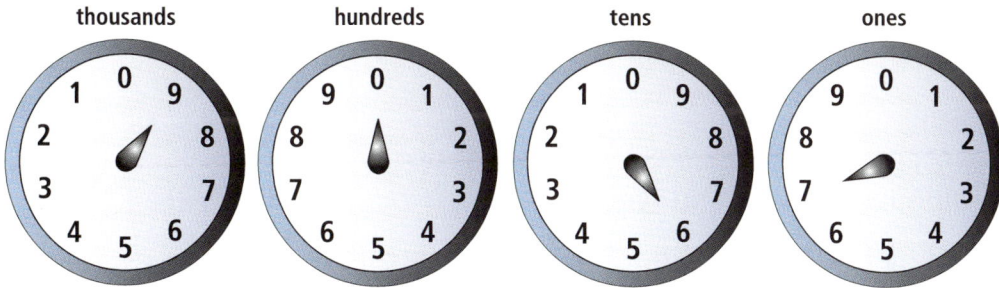

- Always read a meter from left to right. The first dial on the left records thousands of kilowatt hours. The second dial records hundreds of kilowatt hours. The third dial records tens of kilowatt hours. The fourth dial records single kilowatt hours.
- When a pointer is between two numbers, record the smaller number.

Lesson 7: Multiplying by 10, 100, and 1000   **333**

- If your meter says "MULT x 10," the first dial records ten thousands of kilowatt hours, the second dial records thousands of kilowatt hours, the third dial records hundreds of kilowatt hours, and the fourth dial records tens of kilowatt hours.

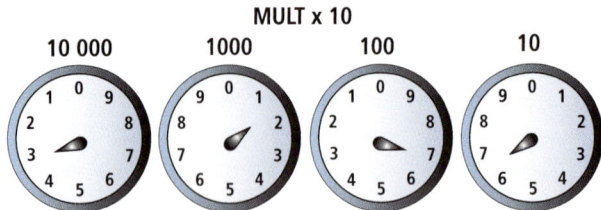

Here's an example. The dials in this picture read 3, 1, 7, and 6. If we multiply by 10 to find the number of kilowatt hours (kWh), this meter reads 31 760 kWh.

Journal
Write 31 760 in your journal in pictures and words.

1. Work with a partner to read each electric meter above. Write each meter reading in numbers and words.

2. Imagine that each meter dial has the following readings:

   a) 3   b) 4   c) 5   d) 2

   Multiply each one by 10, 100, and 1000.

3. What are the patterns that you notice? What rule can you state when multiplying by 10, 100, and 1000?

Chapter 7: Electric Math

4. Now multiply 3452 by 10. Use your rule to predict what the product would be if you were to multiply the same number by 100 and 1000. Use a calculator to see if your predictions are correct.

Build Your Understanding

Read Meters

A meter is a kind of chart. Practise reading the meters below.

1. In your notebook, make a chart like the one below.

Meter Letter	Dial 1	Dial 2	Dial 3	Dial 4	Reading	MULT x 10	Write the Number in Words
A							
B							
C							
D							
E							
F							
G							
H							

Lesson 7: Multiplying by 10, 100, and 1000

2. Record your readings on the chart. There are eight sets of dials in the illustration on the previous page, so you will have eight readings.

3. Compare your answers with another classmate's answers. If you disagree on the readings, check them again.

4. Order and record the readings from greatest to least.

5. What is the mean reading?

6. What is the difference between the greatest and least reading?

7. Make up three more math problems based on the readings. Share them with your partner. Answer the three problems your partner has made up.

What Did You Learn?

1. Use your own words to write the directions for reading a meter. Use your directions to teach someone who does not know how to read a meter. How well did he or she understand your directions? How could your directions be clearer?

2. You have been working with large numbers. What did you discover about multiplying a number by 10, 100, and 1000? Use examples to support your answer.

Practice

Extension

1. **a)** Find out where the meter is located in your school. Your teacher will take your class to read the meter on a regular basis for the next few weeks.

   b) Choose an exact time and day to read your meter. For example, read it every day at 10:00 A.M.

   c) Record the number on each dial along with the date and time.

   d) Subtract the first reading from the second to find out how much energy your school used since the last reading.

   e) Record your data on a class chart.

2. a) Write a rule for dividing by 10, 100, and 1000. Use what you have learned about multiplying by 10, 100, and 1000 to help you.

   b) Share your rule with a partner.

   c) With your partner, take turns dividing large numbers by 10, 100, and 1000, using your rule to predict the answers.

   d) Check your predictions with a calculator.

Lesson 8

Looking at Energy Tables

ENERGY NOTE

PLAN:
You will examine and analyze tables about electricity use over a year.

DESCRIPTION:
It is important to know how much electricity is used from year to year. The Mathford Electric Company knows this and so do the people who receive the electricity bills.

Get Started

Every month a meter reader visits your home to read the dials on your electric meter. The electric company subtracts last month's reading from this month's reading. The difference is the number of kilowatt hours your family used in the last month.

The number of kilowatt hours you use determines how much your electricity bill is. The electric company multiplies the number of kilowatt hours you use by the dollar amount that it charges for one kilowatt hour.

In some households, the meter is read in alternate months and only estimated in the other months.

1. In a month where the meter was read, the kilowatt hours used were 861 kWh, while the estimate was 918 kWh. What is the difference in actual usage compared to estimated usage?

2. What would be the annual difference in usage if this difference occurred 6 times per year? Show your calculations.

Chapter 7: Electric Math

Build Your Understanding

Analyze the Energy Table

This is an Energy Consumption History table for one household. This information allows customers to see how much electricity they use in a year.

Journal

In your journal, discuss how you used the table and graphs to discover information about this consumer's energy consumption.

Your Account Number: 001			
Date Read	Days	kWh	kWh/day
July 10, 2003	33	540	16
June 7, 2003	31	440	14
May 7, 2003	33	520	16
April 4, 2003	28	400	14
March 7, 2003	29	420	14
February 6, 2003	33	330	10
January 4, 2003	30	400	13
December 5, 2002	29	370	13
November 6, 2002	26	360	14
October 6, 2002	40	630	16
September 12, 2002	22	300	14
August 13, 2002	30	420	14

Work with a partner to answer the questions below.

1. What is the total amount of energy used for all months?

2. What is the mean amount of energy used per month?

3. Make a bar graph. Display the monthly use of kilowatt hours for this household for one year. Remember to title and label your graph.

4. Make another bar graph with four bars—one for each season. For example, you could group December to February as winter. Calculate the mean amount of energy used per season. Use a calculator to help you. What can you learn from this graph?

Lesson 8: Looking at Energy Tables

What Did You Learn?

1. What have you learned about this household's use of electricity?

2. This home has air conditioning. When would air conditioning be used? How would it affect energy consumption? Did it? Explain your answer using the data.

3. In 10 years, how much electricity will have been used if it continues to be used at the same rate? Explain how you calculated your answer.

Practice

Find the total for each set of numbers in two ways—starting at the beginning of the set and at the end. Check that the two totals are the same.

1. 5, 7, 3, 4, 5, 7, 2, 6, 9, 8, 2, 6, 1, 4, 3
2. 8, 8, 4, 5, 5, 7, 9, 2, 7, 3, 3, 7, 5, 8, 4
3. 5, 4, 9, 1, 5, 6, 7, 3, 2, 2, 5, 6, 3, 1, 7
4. Find the mode for each set of data in questions 1 to 3.
5. Find the mean for each set of data in questions 1 to 3. Round each number to the nearest whole number.

Solve the following:

6. 0.43 + 0.51 = ▪
7. 7.87 − 5.2 = ▪
8. 13.54 + 5.25 = ▪
9. 0.091 × 10 = ▪
10. 623 ÷ 10 = ▪
11. 0.005 × 1000 = ▪

Extension

12. **a)** Energy costs for this household are 12¢ per kWh. Calculate how much the family paid during the 12 months recorded in the table. You may use a calculator.

   b) Calculate the daily electricity cost.

   c) Calculate the mean daily kilowatt-hour cost.

DATA MANAGEMENT AND PROBABILITY
NUMBER SENSE AND NUMERATION

Lesson 9
Comparing Data

ENERGY NOTE

PLAN:
You will compare two tables and two graphs to see which house uses less electricity.

DESCRIPTION:
Every household uses different amounts of energy depending on how many people are living in the home and using energy.

Get Started

With your class, discuss why there might be differences in how much electricity different households consume. For example, would a family of three use more electricity than a person living alone? Why or why not?

Build Your Understanding

Compare Tables

In Lesson 8, you analyzed the Energy Consumption History table for Account 001. The next page shows an Energy Consumption History table for Account 002. Account 001 is a home with two parents and one child, and Account 002 is a home with one person.

1. Compare the information in this table to the one in Lesson 8.

2. Make a bar graph. Display the monthly use of kWh for Account 002 for one year.

Lesson 9: Comparing Data   341

Your Account Number: 002			
Date Read	Days	kWh	kWh/day
July 6, 2001	30	415	14
June 6, 2001	32	329	10
May 8, 2001	29	311	11
April 9, 2001	34	416	12
March 7, 2001	27	354	13
February 8, 2001	30	433	14
January 9, 2001	30	553	18
December 7, 2000	31	586	19
November 9, 2000	32	384	12
October 4, 2000	32	378	12
September 6, 2000	30	417	14
August 2, 2000	34	408	12

3. Compare your bar graph to the graph you made in Lesson 8. Which household consumed more energy over the 12 months?

4. Calculate how many kilowatt hours Account 002 used during the 12 months. What is the mean amount of energy consumed per month? Compare this with Account 001.

5. Write three other sentences to describe how the two bills are the same or different.

What Did You Learn?

1. What did you find when you compared the information in the two tables?

2. How did you calculate the mean monthly energy use?

Practice

Write these numbers in your notebook from least to greatest. Use the greater than sign (>) or less than sign (<) between every pair of numbers. The first one has been done for you.

1. 56 321, 493 300, 231 402, 561 230, 4123

   4123 < 56 321 < 231 402 < 493 300 < 561 230

2. 32 316, 23 316, 4518, 4581, 62, 499 823

3. 450 321, 432 100, 987 654, 992 323, 512

4. 7456, 32 311, 40 581, 671 003, 671 118, 3270

5. What strategy or method do you use for ordering a set of numbers from least to greatest? Give examples to support your answer.

Extension

6. **a)** As a class, look at actual electricity bills. Check if there is an Energy Consumption History table, or a similar table, showing electricity use over the past year. Find out the mean monthly use of electricity as shown on the bills.

   b) Make up math questions based on the actual electricity bills. Take turns asking questions to the class. Make sure you have the answers prepared for your questions.

Technology

Different cities and districts charge different amounts per kWh of electricity usage. Use the Internet and other resources to compare current electricity rates in your area with those in other locations.

Lesson 10
Calculating Costs

ENERGY NOTE

PLAN:
You will calculate the cost of electricity.

DESCRIPTION:
Using less electricity can save the people of Mathford money. Find the cost of a month's supply of electricity.

Get Started

With your classmates, imagine that these are the current provincial and territorial electricity prices.

Hydro Prices (Cost per kWh)			
British Columbia	$0.07	Nova Scotia	$0.10
Alberta	$0.07	Prince Edward Island	$0.12
Saskatchewan	$0.08	Newfoundland and Labrador	$0.09
Manitoba	$0.06	Northwest Territories	$0.12
Ontario	$0.10	Yukon Territory	$0.11
Québec	$0.07	Nunavut	$0.32
New Brunswick	$0.08		

1. Where is electricity the most expensive?
2. Where is it the least expensive?
3. Arrange the prices in order from the lowest to the highest. Where in the list does your province or territory fall?
4. Calculate the national mean. Show your work.
5. What is the mode? How do you know?

Chapter 7: Electric Math

Build Your Understanding

Calculate Your Bill

Work in a small group. Calculate the electricity bills for four Canadian homes. Use the information in the chart and map to help you.

1. First, locate each household on the map.

2. Look at the chart of the 10 provinces and 3 territories on the previous page, and find the electricity cost for each household. State a formula that you could use. You may use a calculator to help you.

3. Make a chart in your notebook to record the four totals.

Household	kWh Used Last Month
1	540
2	440
3	520
4	400

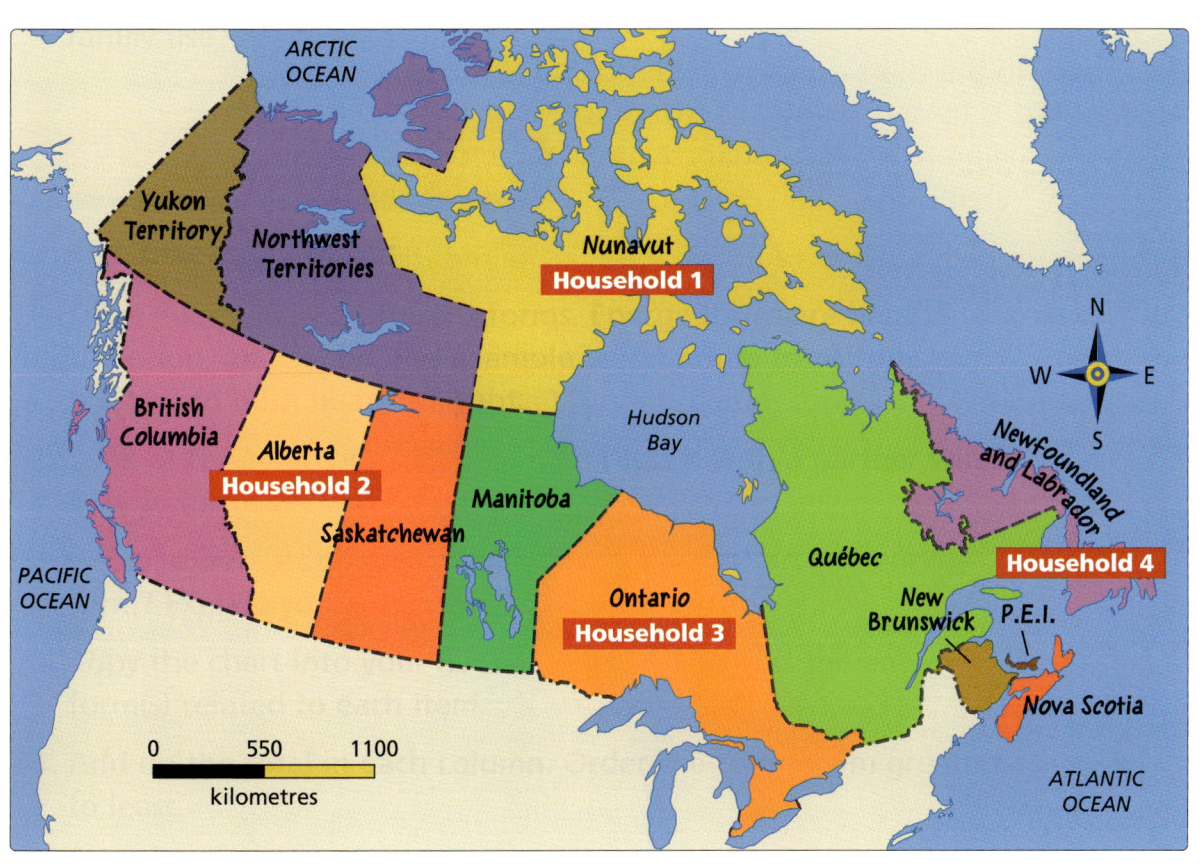

Lesson 10: Calculating Costs 345

What Did You Learn?

1. Put the households in order from the lowest number of kilowatt hours used to highest.

2. Put the households in order from lowest electricity bill to highest.

3. Compare the two lists. Explain the differences.

4. Calculate the cost of the mean bill.

5. Calculate how much each household is above or below the mean bill. Show your work.

6. Compare your answers with those of another classmate.

Practice

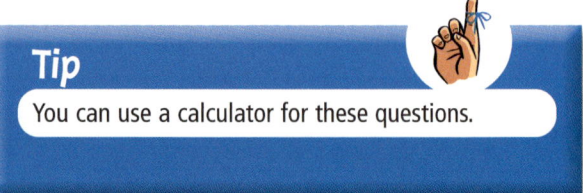

Tip
You can use a calculator for these questions.

Extension

Use pictures, numbers, and words to answer the following questions.

1. Refer back to the chart on page 342 in Lesson 9. Suppose that Account 002 is for a household in the Northwest Territories. If the month being billed is February, what is the monthly cost for electricity? What is the daily mean cost?

2. If the month being billed is August, what is the monthly cost for electricity? What is the daily mean cost?

3. Refer to the chart on the previous page. Imagine all four households are from your neighbourhood. Calculate their electricity bills for the last month using your provincial rate.

Chapter 7: Electric Math

Show What You Know

Review: Lessons 7 to 10, Multiplication and Calculating Costs

1. If a meter reading was 2355 and you wanted to calculate the kilowatt hours used, what would you do? Explain how you would figure this out in pictures, numbers, and words.

2. Use pictures, numbers, and words to answer these questions:

   a) If one family used 12 kWh for 29 days and the cost was $0.11 per kWh used, calculate the energy cost.

   b) If another family used 11 kWh per day for 33 days and the cost was $0.10 per kWh used, calculate the energy cost.

   c) Compare the two costs in 2 (a) and (b). What do you notice? Share your comparison with a classmate.

Lesson 11
Finding Perimeter and Area

ENERGY NOTE

PLAN:
Find the perimeter and area of the meter reader's routes, which will form the shapes of irregular polygons.

DESCRIPTION:
A Mathford meter reader does a lot of walking. Follow him or her around and see how far he or she walks.

Get Started

If the meter reader walks 7 h a day at an average pace of 3 km/h, how many kilometres does he or she walk in one day? If he or she increases his or her speed to 4 km/h, how many more kilometres does he or she cover?

Build Your Understanding

Calculate Area

You Will Need
• grid paper

This meter reader has two routes. Each route takes one week to cover. The meter reader works for two weeks and then has two weeks off.

Part 1

Plot each route on the grid paper.

1. On your grid paper, label the *x*-axis from A to Q.
2. Label the *y*-axis from 0 to 18. (Remember to label lines and not spaces.)

Chapter 7: Electric Math

3. Plot the following data. Start at (A, 11), which is where the electric company's office is located.

Route 1: (A, 11) to (C, 1) to (O, 1) to (N, 8) to (K, 6) to (I, 7) to (F, 7) to (E, 4) to (A, 11).

Route 2: (A, 11) to (F, 8) to (H, 9) to (H, 14) to (L, 13) to (L, 10) to (P, 11) to (Q, 16) to (F, 17) to (E, 12) to (A, 11).

Part 2

Find the perimeter and area of each route.

1. If each square on the graph represents one square kilometre, how many kilometres does the meter reader walk in one year? (Remember, the meter reader works two weeks on and two weeks off.) Use pictures, numbers, and words.

2. Estimate the area that each route covers.

Journal
Record your answers in your journal before sharing them with the class.

Tip
To find the perimeter, add up the length of each segment of the route. The side of each square on the graph represents one kilometre.

What Did You Learn?

1. What problem-solving strategies did you use to find the area of the meter reader's routes?

2. These routes form irregular polygons. Explain to a partner how you estimated the area and calculated the perimeter of these irregular polygons.

Practice

Extension

Add the following:

1. 987 + 560 + 125 + 294 = ■

2. 21 590 + 36 897 = ■

3. 3194 + 3015 + 6218 = ■

Lesson 11: Finding Perimeter and Area

Subtract the following:

4. 3956 − 1131 = ▨

5. 4700 − 1900 = ▨

6. 5615 − 1400 = ▨

Multiply the following:

7. 48 x 4 = ▨   **8.** 75 x 4 = ▨

9. 420 x 4 = ▨   **10.** 62 x 23 = ▨

11. 41 x 32 = ▨ **12.** 31 x 99 = ▨

13. a) Design more problems for your classmates to solve. Here's an example: The meter reader has a new route. Record the coordinates of the new route.

   b) Plot the new route on grid paper.

   c) Estimate the area of the new route.

14. a) If the meter reader reads 110 meters on Day 1, 117 meters on Day 2, 122 meters on Day 3, and 129 meters on Day 4, how many meters will she read on Day 7? on Day 10?

   b) How many meters in total will she have read by the end of Day 15?

   c) On what day will she read the one thousandth meter?

   d) Record your answers in chart form. What patterns do you see? State the rule.

15. If the meter reader reads 1000 meters in a week, and works 5 days a week, what fraction of meters would she have read after 2 days? Explain your answer using pictures, numbers, and words.

Technology

Use a paint/draw computer program to create a floor plan of your home. Label the rooms and include the measurements of each. Calculate the perimeter and area of each of the rooms, using a spreadsheet computer program.

Chapter 7: Electric Math

Chapter Review

1. Find the mean of each set of numbers:
   a) 32   40   97   82
   b) 960   946   499   600   329
   c) 1 234   2 456   3 009   1 765

2. Find the mode of each set of numbers:
   a) 7   6   2   1   7   7   1
   b) 34   92   34   61   92   40

3. Order each group of numbers from greatest to least:
   a) 56 325, 39 402, 69 301, 89 222, 1258
   b) 33 258, 36 415, 31 598, 39 856, 34 593
   c) 65 298, 65 012, 65 349, 65 987, 65 556
   d) 78 549, 78 503, 78 529, 78 598, 78 513

4. Mark finished his homework in $4\frac{1}{2}$ h sitting under an overhead light that uses one 40 W light bulb. Karissa finished her homework in 3 h sitting beside a table lamp that uses one 60 W light bulb. Who used more electricity? Use pictures, numbers, and words.

5. Imagine that two DVD players are on sale for (1) $199.00 and (2) $245.00. The first DVD player uses 400 kWh of energy and the second one uses 350 kWh. Both DVD players have a life expectancy of 5 years. Which DVD player is the better buy? Use pictures, numbers, and words.

6. Your family wins a coupon for $100.00 to spend on small electrical appliances. You may buy only from the list in the table. List as many different combinations as possible without going over $100.00. The prices below include taxes.

Curling Iron:	$13.74	Iron:	$30.00	Electric Shaver:	$27.99
Electric Frying Pan:	$42.29	Coffee Maker:	$59.99	Toaster:	$32.50
Electric Can Opener:	$5.69	Clock Radio:	$24.62	Hair Dryer:	$24.99

7. Multiply these numbers by 10, 100, and 1000:

	X 10	X 100	X 1000
(a) 8617			
(b) 24 203			
(c) 98 069			

8. Divide these numbers by 10, 100, and 1000:

	10	100	1000
(a) 5613			
(b) 75 420			
(c) 89 064			

9. Look at the information in this chart showing the number of appliances used in one week.

Appliances Used	
Thomas	3
Tavia	7
Kenny	4
Dan	11

   a) Write the fraction of appliances used by each person.
   b) Make a bar graph showing the information.
   c) Calculate the mode and mean.
   d) Write three questions about the data for a classmate to answer.

10. If one household used 15 kWh for 30 days and the cost was $0.09 per kWh used, calculate the energy cost. Use pictures, numbers, and words.

11. On a piece of grid paper, draw an irregular shape. Calculate the perimeter and area of your shape.

Chapter 7 Chapter Wrap-Up

DATA MANAGEMENT AND PROBABILITY
NUMBER SENSE AND NUMERATION
MEASUREMENT
GEOMETRY AND SPATIAL SENSE

In this chapter, you have learned how to measure and calculate the cost of the electricity we use daily.

Imagine that you work for the Mathford Electric Company and are analyzing electricity use.

Here is the data for one family for 12 months. Imagine that the rate is $0.08/kWh.

Energy Consumption History (This Year)					
Month	Amount Billed	kWh	Month	Amount Billed	kWh
Jan.	$110.80	1385	July	$30.40	380
Feb.	$87.20	1090	Aug.	$38.40	480
March	$79.20	990	Sept.	$132.00	1650
April	$62.80	785	Oct.	$83.60	1045
May	$30.40	380	Nov.	$85.60	1070
June	$28.80	360	Dec.		1200

1. Calculate the family's payment for December. Use a calculator if you need to.
2. Find the mean amount of money spent on electricity over the year. Show your work.
3. Find the mean amount of electricity used.
4. Graph the year's use of kilowatt hours.
5. Write four observations you can make from the data.
6. When is this household's electricity use highest? Suggest three ways the family might lower its use.

7. You are going to compare the electricity costs from this year to the electricity costs from last year. Here is the data from last year. Imagine that the rate is $0.08/kWh.

Energy Consumption History (Last Year)					
Month	Amount Billed	kWh	Month	Amount Billed	kWh
Jan.	$103.20	1290	July	$32.80	410
Feb.		1055	Aug.		460
March	$82.00	1025	Sept.		1600
April	$64.00	800	Oct.	$84.40	1055
May		375	Nov.		1170
June		365	Dec.		1300

8. Copy and complete the chart in your notebook.

9. Write three statements to compare the data in last year's and this year's charts. Share your statements with a classmate.

10. a) Create some math questions that could be answered from the data in the chart above. Make sure you have at least one addition, one subtraction, and one multiplication question.

   b) Calculate the answers to your questions.

Chapter 8

Seedling Sums

An important part of conserving energy is conserving the resources that provide us with energy. Trees help us save energy by providing shade to keep us cool in the summer. We need to conserve trees too.

In this chapter, you will

- examine fractions, costs, and patterns about trees
- calculate volume and capacity
- research, record, analyze, order, and compare data
- create tessellating patterns
- find the mean and mode for data

At the end of this chapter, you will analyze, evaluate, and graph data on the hourly energy consumption for Mathford.

1. What are the similarities and differences between bar graphs and line graphs?
2. In your journal, explain volume and capacity using pictures, numbers, and words.

NUMBER SENSE AND NUMERATION
PATTERNING AND ALGEBRA

Lesson 1

Calculating Tree Costs

ENERGY NOTE

PLAN:
You will calculate how many trees will be planted according to different plans. You will also calculate the costs of planting trees.

DESCRIPTION:
It gets very hot in Mathford during the summer, and the citizens want to know how to stay cool without using a lot of energy. The shade from trees provides one way to stay cool without using electricity.

Get Started

In the summer, sometimes the temperature can go as high as 30°C. That's hot! One way to keep cool when you are outdoors is to sit under a shady tree.

Some people have air conditioners to cool their homes. One way to keep your home cool that doesn't use electricity is to surround it with lots of shady trees.

1. Imagine that you planted 10 maple trees at a total cost of $49.99. What would be the cost of each tree?

2. Based on the cost of each tree that you calculated in question 1, how much would it cost to plant 100 and 1000 trees?

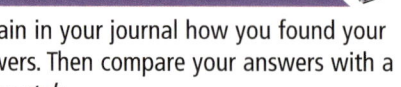

Journal

Explain in your journal how you found your answers. Then compare your answers with a classmate's.

Chapter 8: Seedling Sums

Build Your Understanding

Multiply Trees

Your town has decided to plant more shady trees. The town council has been presented with two plans:

- Plan A recommends planting 1750 trees this year.
- Plan B recommends planting 25 trees this year, 37 trees in the second year, 50 trees in the third year, 64 trees in the fourth year, and so on, for 10 years.

1. Estimate which plan would produce the most trees. Record your estimates for Plan A and Plan B.

2. What is the pattern in Plan B? Make a table and show how many trees would be planted after 10 years. How many trees would be planted by the end of the fourth year? the eighth year? the tenth year?

3. Would more trees be planted with Plan A or Plan B?

4. What is the difference between the number of trees planted with Plan A and with Plan B?

5. If each tree costs the town $150.00 in the first year, how much would Plan A cost? Show your work.

6. If the cost for a tree increases by $3.00 every year after the first year, how much would Plan B cost? Show your work.

What Did You Learn?

1. Study the table you created. Based on costs, which plan would you recommend to the town council?

2. Meet in a small group to share, compare, and discuss your recommendations.

Practice

Pretend the town council didn't like Plan A or Plan B. Create Plan C using a different pattern. Identify your pattern. Share your plan with the class.

Lesson 1: Calculating Tree Costs

Lesson 2

Fractions of Whole Numbers

ENERGY NOTE

PLAN:
You will find fractions of whole numbers of trees.

DESCRIPTION:
Many trees on the outskirts of Mathford were cut down last year. Now the town council wants to plant new seedlings, because they know land needs replanting after trees have been cut down.

Get Started

1. With your classmates, look at the illustration above of tree seedlings that have just been planted. How many is $\frac{1}{10}$ of the seedlings in the picture? How did you get your answer?

Chapter 8: Seedling Sums

2. Explain how to find a fraction of a whole number using pictures, numbers, and words. Share your explanation with the class.

3. Practise finding more fractions of whole numbers. How much is $\frac{1}{10}$ of 40? of 60? of 100?

4. Make a table to show your data. What patterns do you see? Describe what you discovered.

Tip
You may discover that you are actually dividing the whole numbers by 10. Try your discovery out on a few more numbers: 110, 120, 150.

Technology
Use a paint/draw computer program to show different fractions of a whole number. You can use shapes to show fractions as part of a set, or divided objects to show fractions as part of a whole. Challenge a classmate to write the fractions.

5. Write a rule for dividing by 10 and share it with your class. What is $\frac{1}{10}$ of 1000?

6. Now find $\frac{2}{10}$ of 40, 60, and 100. Use your table to organize your data. What is $\frac{3}{10}$ of 100?

7. Find $\frac{1}{2}$ of 10, then $\frac{1}{5}$, $\frac{2}{5}$, $\frac{3}{5}$, and so on. What is $\frac{2}{5}$ of 100?

8. Find $\frac{1}{2}$ of the trees shown. Then find $\frac{1}{5}$, $\frac{2}{5}$, and $\frac{3}{5}$. Show your answers in pictures, numbers, and words.

Lesson 2: Fractions of Whole Numbers

Build Your Understanding

Calculate Fractions

You Will Need
- hundreds chart
- coloured pencils

You have been hired by the town of Mathford to plant tree seedlings in the far North during the summer. In your first 5 days on the job, you planted 1000 seedlings. The fifth day was cut short because of lightning and thunderstorms. Here's a table to show what you did.

Day	Weather Conditions	Fraction of Seedlings Planted	How Many Seedlings Were Planted?
1	sunny	$\frac{1}{5}$ of 1000	
2	cloudy	$\frac{1}{4}$ of 1000	
3	rainy	$\frac{1}{10}$ of 1000	
4	sunny	$\frac{2}{5}$ of 1000	
5	thunderstorms	You planted the rest of the 1000 seedlings.	

Work with a classmate. Use a hundreds grid like the one shown to the right and the data from the chart to answer these questions. Each square represents 10 seedlings.

1. Copy the table above into your notebook. Calculate how many seedlings were planted each day. Explain how you got your answer.

2. Show your calculations from question 1 on a hundreds grid. Use a different coloured pencil to represent each day. Make a legend to go with your grid.

3. Now that you know how many seedlings each fraction represents, order the fractions from least to greatest.

Chapter 8: Seedling Sums

4. On which day were the most seedlings planted? On which day were the fewest seedlings planted? What is the mean number of seedlings planted per day?

5. If the weather continues to be warm and sunny, and you plant as many seedlings each day as you did on Day 4, how many seedlings will you plant in the next 2 weeks? Show your work using pictures, numbers, and words.

6. How many seedlings will you have planted altogether?

What Did You Learn?

1. Check your answers to the Build Your Understanding questions with another pair of students.

2. How did the hundreds grid help you answer the questions?

3. Write a rule for finding a fraction of a whole number, such as $\frac{1}{2}$ of 8 or $\frac{2}{3}$ of 9. Teach someone your rule. Does it work every time? Explain why or why not.

Practice

Extension

We can also analyze the data from the table using ratios. A ratio is a number compared to another number. We use the symbol ":" to express ratios.
To show that it was sunny on 2 out of the 5 days, the ratio would be 2:5. To show that it was cloudy on 1 out of the 5 days, the ratio would be 1:5.

1. How would you write the ratio to show that 3 out of 5 days had no rain?

2. What does the ratio 600:1000 mean? What does the ratio 150:1000 mean?

A Math Problem to Solve

3. In another area, you worked with a planting crew to plant 2500 seedlings in 2 weeks. Create a chart showing what fraction of seeds you planted on each working day, remembering that you worked 5 days a week. How many trees did you plant on each working day? Your chart needs to account for weather changes and for 2 days when one crew member was absent. Use pictures, numbers, and words.

Lesson 2: Fractions of Whole Numbers

Lesson 3

Calculating Your Earnings

ENERGY NOTE

PLAN:
You will calculate tree-planting earnings.

DESCRIPTION:
You can make a lot of money planting trees for the town of Mathford. Find out how much you can earn.

Get Started

1. If you plant 350 trees a week, how many trees will you plant in 1 day, assuming you work 5 days a week? If you increase your planting speed by 5 trees a day, how many more trees will you plant in a week? Share your answers with your classmates.

2. Answer the following questions:
   a) 256 ÷ 3
   b) 4) 2546
   c) 3125 ÷ 4
   d) $0.36 x 4
   e) $1.89 x 6
   f) $4.67
      x 3

Build Your Understanding

Imagine that you have spent 20 days planting trees this summer.

1. If you planted 350 trees a week, working 5 days a week, how many trees did you plant in 20 working days? Compare your answer with that of another classmate.

2. If you were paid $0.25 for each tree that you planted, how much money would you earn in one day? You may use a calculator to help you.

Chapter 8: Seedling Sums

3. Since you worked so hard and didn't complain about the mosquitoes, the town of Mathford has decided to pay you $0.50 for each seedling that you planted during the last 5 days of planting. How much money will you make during these days? You may use a calculator to help you.

4. How much will you now have earned in all 20 days of planting? Show your work.

What Did You Learn?

1. Switch problems with a classmate and check each other's solutions.

2. Compare your answers and discuss the problem-solving strategies that each of you used.

3. Did you solve the problems in the same way? What was different about how you each solved the problems?

Practice

Extension

1. If you worked the entire summer, how much money would you make? What do you need to know to answer this question? Use pictures, numbers, and words to show your answers.

2. Create a word problem about planting trees like the ones in this lesson. Write a solution to your problem. Give your problem to a classmate to solve. Compare your answers and problem-solving strategies.

Lesson 3: Calculating Your Earnings

Lesson 4
Transformations, Fractions, and Patterns

ENERGY NOTE

PLAN:
When tree planting, you will use patterns to show one half of a whole, and you will calculate fractions of a whole. You will also study translations, rotations, and reflections.

DESCRIPTION:
Trees are planted in grid patterns with each planter working in a different square. This way, an even number of trees are planted in the entire area.

Get Started

You Will Need
- grid paper

Fold a sheet of grid paper in half and then in half again to create four equal sections. In one section, draw a pattern of a leaf. Use the other three sections to show what the leaf would look like after it is reflected (flipped), translated (slid), and rotated (turned). Describe your transformations in as much detail as possible, using mathematical terms. Share your answers with a classmate. Why might your answers look different?

Vocabulary

axis: One of the intersecting number lines on a graph. Two or more of these lines are called axes.
hectare: A unit of measurement equal to one square hectometre. A hectometre equals 10 000 m.
reflection (flip): A transformation (movement) of a figure by flipping it over a mirror line
rotation (turn): A transformation (movement) of a figure around a fixed point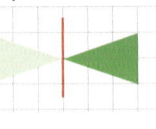

transformation: A movement that does not change the size or shape of a figure
translation (slide): A transformation (movement) of a figure along a straight line

Build Your Understanding

Calculate Half of a Whole

You Will Need
- letter-sized paper
- coloured pencils

The piece of land where you are planting trees has been staked out in hectares. Each hectare forms a square. Your planting crew has been asked to plant only half of every hectare to allow for future tree growth.

1. **a)** Take a piece of letter-sized paper and fold over the top right corner so its point is touching the left side.

   b) Cut off the bottom rectangle of paper and recycle it. You should now have a square. Let your square represent one hectare of land.

2. Each hectare must be divided into eighths ($\frac{1}{8}$). Show the different ways you can divide your paper into eighths.

3. On your hectare, use different coloured pencils to shade in one way to plant half of a hectare. Use other pieces of square paper to show different ways to plant half of a hectare. Do not repeat any patterns.

4. Check that your planting patterns are all different and that none of them are actually patterns you have already used that have been reflected, translated, or rotated.

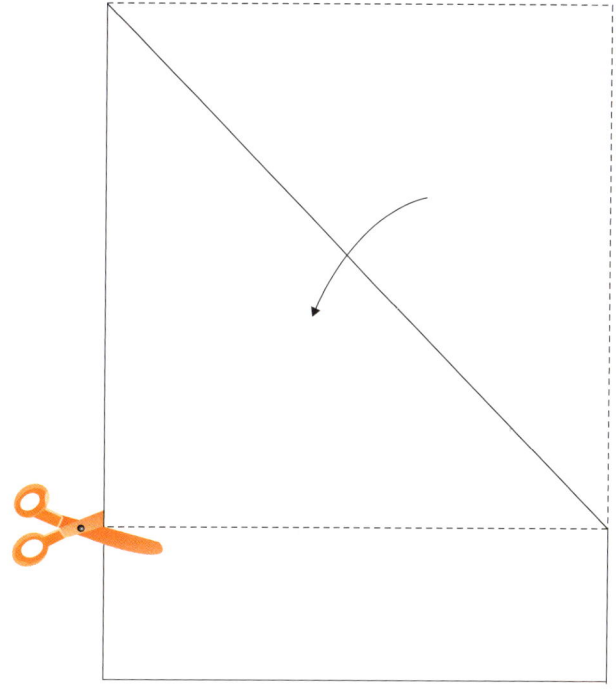

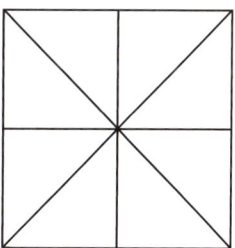

Lesson 4: Transformations, Fractions, and Patterns

What Did You Learn?

1. Mount each square that you made in Build Your Understanding on a piece of Bristol board. How many different patterns did you find?

2. Make a table to show how many different patterns are possible for a square divided into halves, quarters, and eighths. What do you notice?

3. With a classmate, predict how many patterns you would find for a square divided into sixteenths.

Journal

In your journal, explain how you determined if the patterns were different. Use pictures, numbers, and words to record your explanation.

Practice

1. On Monday, you planted enough trees to cover $\frac{1}{8}$ of a hectare, on Tuesday $\frac{1}{2}$, on Wednesday $\frac{1}{4}$, and on Thursday $\frac{1}{3}$. On which day did you cover the greatest area? On which day did you cover the least? Use pictures, numbers, and words.

2. On Friday, you planted enough trees to cover $1\frac{1}{8}$ hectares. If you planted 50 trees in every $\frac{1}{8}$ hectare, how many trees did you plant?

3. Draw lines of symmetry through the patterns that you coloured in Build Your Understanding.

4. If one tree needs $\frac{1}{16}$ of a hectare of room, how many trees can you plant in 5 hectares? Explain how you got your answer.

Technology

Using a paint/draw computer program, create a chart using the line tool. Add the headings Object, Translation, Reflection, and Rotation. Create different objects in the first column. Apply the appropriate transformation to the objects to complete the rest of the chart.

Vocabulary

symmetry: Two congruent sides divided by a line

Show What You Know

Review: Lessons 1 to 4, Cost, Fractions, and Geometry

Imagine that you planted 500 seedlings in 4 days on 1 hectare of land.

1. If you planted the same number of seedlings each day, what fraction would you have planted each day?

2. If you gave each seedling 1 m^2 of space, how many square metres would be required for all of the seedlings?

3. Would 1 hectare be enough to plant 500 seedlings? Explain why or why not.

4. If you were paid $0.45 per seedling planted the first day, $0.20 per seedling planted the second day, $0.25 per seedling planted the third day, and $0.27 per seedling planted the fourth day, how much money would you make? Use pictures, numbers, and words.

5. a) Using grid paper, create four equal sections by folding your paper in half and then in half again. In one section, draw one of the seedlings that you planted. Use the other three sections to show what the seedling would look like after it was reflected, translated, and rotated.

   b) Describe your transformations using math words. Be as specific as you can.

6. On a different piece of grid paper folded into four equal sections, draw the same seedling you drew in question 5(a), in the same section on your paper. Give your paper to a classmate. Read the description you wrote of your transformations in question 5(b) for your partner to follow. See if your partner's transformations turn out the same as yours. If they are different, how might you change your description to make it clearer?

GEOMETRY AND SPATIAL SENSE

Lesson 5

Water Volume and Capacity

ENERGY NOTE

PLAN:
You will calculate the volume and capacity of different shapes and measure the amount of water used.

DESCRIPTION:
Trees need water to grow, especially just after they have been planted.

Get Started

What is the volume of this box of seedlings? How do you know?

Explain to a classmate how you found the answer.

Build Your Understanding

Water Capacity

Seedlings need water to grow. Those planted in the wilderness must rely on the rain, but if you plant trees in your backyard, you can help them by watering regularly. Work with a classmate to solve these problems. Show your answers in pictures, numbers, and words.

1. Each seedling needs 500 mL of water daily. How many litres (L) is this? What fraction of a 2 L jug is this?

Chapter 8: Seedling Sums

2. If you planted 10 trees today, how many 2 L jugs of water would you need? How many jugs if you planted 50?

3. If you planted 15 trees, how many 2 L jugs would you need?

4. Pretend each tree needs 1250 mL of water, and you plant 3 trees. How many 2 L jugs would you need?

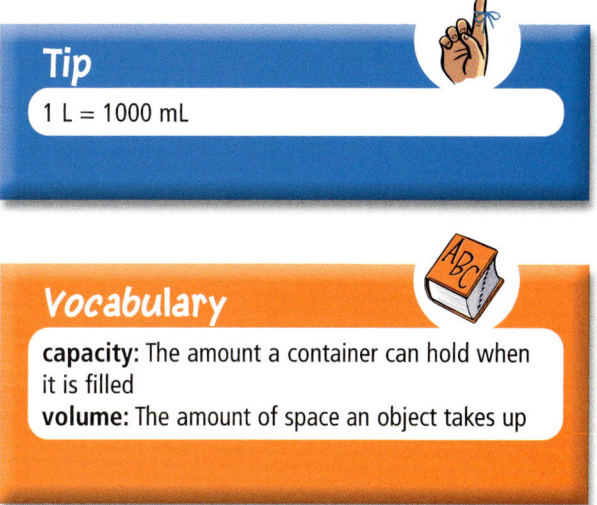

Tip
1 L = 1000 mL

Vocabulary
capacity: The amount a container can hold when it is filled
volume: The amount of space an object takes up

What Did You Learn?

Explain to another pair of students how you solved the problems. Did they solve the problems the same way? What was similar? What was different?

Practice

1. Imagine you are watering trees using a container that measures 5 cm x 8 cm x 10 cm. Calculate the capacity of this container and remember that 1 cm^3 = 1 mL.

2. How much water can a cubic container hold that measures 10 cm on one side? Show your answer in millilitres and litres. Explain how you got your answer.

Convert the following measurements in your notebook:

3. 1 m = ■ mm

4. 34 cm^3 = ■ mL

5. 680 mm = ■ m

6. 48 000 m = ■ km

Tip
1 mL = 1 cm^3

Lesson 5: Water Volume and Capacity

DATA MANAGEMENT AND PROBABILITY
NUMBER SENSE AND NUMERATION

Lesson 6

Collecting and Organizing Paper Data

ENERGY NOTE

PLAN:
You will research, record, and analyze data about paper use.

DESCRIPTION:
Trees are used as a source of energy and for many other things that we use every day. How many trees do we use? The people of Mathford want to know. Did you know that it takes 17 trees to make 1 t (tonne) of paper? North Americans use more than 260 kg of paper a year per person.

Get Started

You Will Need
- calculator (optional)

How would you calculate the amount of paper used by all the students in your class? by your entire school? by your teachers and principal? Share and discuss your answers with the class.

Build Your Understanding

Research and Record Data

Work in a small group.

1. Find out how much paper your class uses in one week. You can measure the number of pieces of paper or the mass of the paper used. Develop a chart to record the amounts of different paper used for the week.

Tip
Think about who you could ask to help you with this information.

2. Create a bar graph of your information.

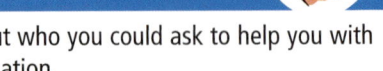
Chapter 8: Seedling Sums

What Did You Learn?

1. How did your bar graph compare to the graphs of the other groups?
2. What conclusions can you draw from your data?

Practice

1. Write problems for other students to solve based on your data. For example, how much paper would your class use in half a year? How much paper would 10 classrooms use in one week?

Extension

2. If your class used $\frac{1}{4}$ less paper, how many pieces or kilograms of paper could be saved in one week?
3. Prepare a presentation for your class, using your research and bar-graph data from this lesson. Be prepared to explain your results.
4. As a class, brainstorm ways to reduce paper use in your classroom.

Technology

Using a word processor, make a list of all the ways you and others use paper. Beside each item on your list, add ways to reduce the amount of paper. Compare your list and ideas with a classmate.

Lesson 7
Data With Fractions and Decimals

ENERGY NOTE

PLAN:
You will order and compare data about heating and work with fractions and decimals.

DESCRIPTION:
Many years ago in Mathford, like everywhere else, people relied on wood for cooking and for heating their homes. Now the people of Mathford have many other ways to heat their homes.

Get Started

It takes energy to heat a home. Wood used to be a common source of energy. Today many homes use natural gas for heating.

Monthly Gas Use in Mathford					
Jan.	Feb.	March	April	May	June
17.89 m^3	15.45 m^3	15.35 m^3	11.64 m^3	5.65 m^3	3.64 m^3
July	Aug.	Sept.	Oct.	Nov.	Dec.
3.24 m^3	3.15 m^3	5.11 m^3	4.25 m^3	6.2 m^3	12.59 m^3

1. Work with a classmate and order these numbers from greatest to least.
2. Calculate the mean.
3. Write each number as a mixed number. (For example, $4.25 = 4\frac{25}{100}$)
4. Read the numbers aloud to your partner. Make sure that you are mathematically correct.

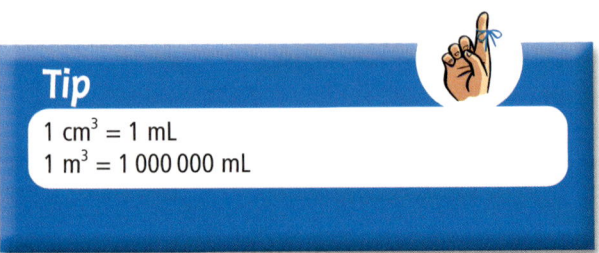

Tip
1 cm^3 = 1 mL
1 m^3 = 1 000 000 mL

Chapter 8: Seedling Sums

5. Write three of the numbers in words, and then ask your partner to check that you wrote them correctly. Do the same for your partner.

6. Round off each number to the nearest tenth.

7. Which was the coldest month in this year? How do you know?

> **Tip**
> Rounding is a rule used to make an approximation to a number. You should round up when the digit is 5 or higher, and round down when the digit is less than 5. For example, if we were to round the first number in the chart on the previous page to the nearest tenth, it would be 17.9 m^3.

Build Your Understanding

Organize Data

1. Work with a classmate. Make charts in your notebook like the ones shown here.

Source of Energy Produced			
Source of Energy	Rounded Off	As a Fraction	In Words

Source of Energy Consumed			
Source of Energy	Rounded Off	As a Fraction	In Words

Lesson 7: Data With Fractions and Decimals

2. Using the data from the tables below, list the sources of energy on the charts you made for question 1. List them in order from greatest to least source of energy produced and from greatest to least source of energy consumed.

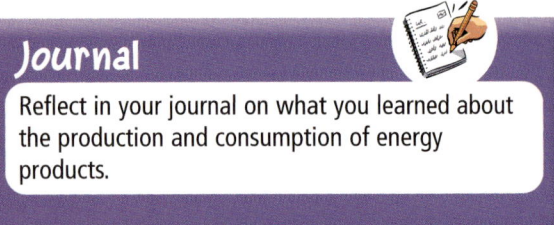

Journal

Reflect in your journal on what you learned about the production and consumption of energy products.

Source of Energy Produced	
Wood	3.80%
Petroleum	35.36%
Gas	38.44%
Coal	9.32%
Electricity	13.27%

Source of Energy Consumed	
Wood	5.68%
Petroleum	37.23%
Gas	26.72%
Coal	11.85%
Electricity	18.74%

3. Round off each number to the nearest tenth.

4. Write each number as a mixed number.

5. Write each number in words.

6. What is the mean of each chart?

What Did You Learn?

1. Compare the production and consumption of each product.

2. Which sources do we consume more than we produce? What is the difference between consumption and production?

3. In your own words, explain how to round a number to the nearest tenth. Use examples to support your explanation.

4. How do you order decimals from greatest to least? Use examples to support your answer.

Practice

Extension

Graph the information presented in the charts. Which type of graph works best? Why?

DATA MANAGEMENT AND PROBABILITY

Lesson 8

Surveys and Graphs

ENERGY NOTE

PLAN:
You will read a graph about energy use.

DESCRIPTION:
The people of Mathford consume all types of energy. Which type do they use the most? Which type do you use the most?

Get Started

With a classmate, work to design a survey of how people heat their homes. You will need to make several decisions with your partner.

1. Create the questions for your survey.
2. Decide on the population you will survey.
    a) How large a group will you survey?
    b) Who will you survey?
3. Conduct your survey and record the responses.
    a) How will you organize your results?
    b) How will you present the data you have collected?

Share your results with the rest of the class. What have you discovered about heating homes in your community?

Lesson 8: Surveys and Graphs   375

Build Your Understanding

Analyze Data

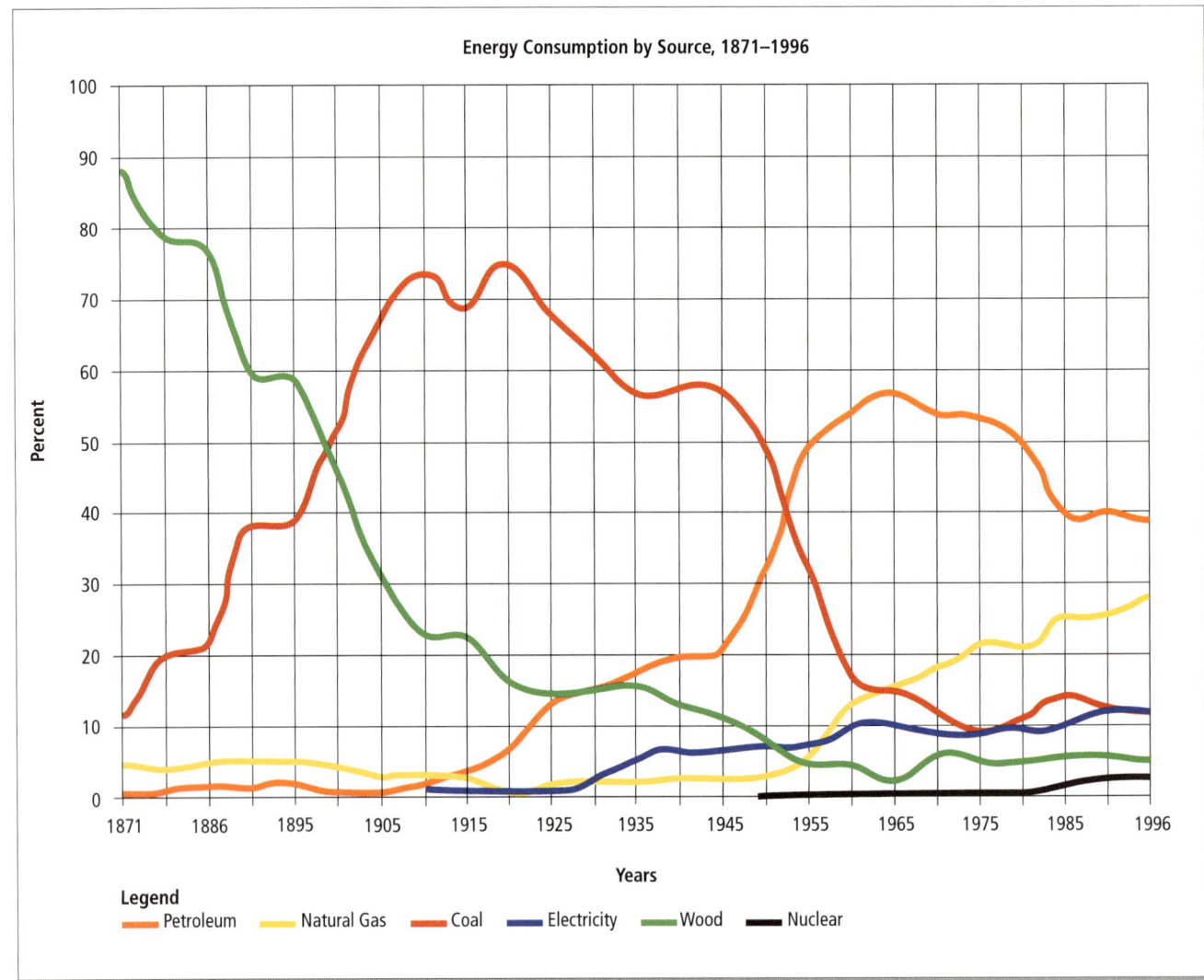

Study this graph from Energy Canada with your classmates. It shows the different types of energy that Canadians used from 1871 to 1996.

1. How many years does this graph span?
2. What energy sources are shown on the graph?
3. What changes in energy use have there been since 1871? What do you think caused these changes?

4. Which source of energy was used most in 1871? Why do you think it was so widely used?

5. Which source of energy was used most in the early 1900s?

6. Which source of energy has consistently been used less and less over the years? Explain why this might be.

7. Which source of energy has steadily been used more and more each year? Explain why this might be.

What Did You Learn?

1. How has the choice of energy use changed over time? What caused these changes?

2. Look back at your survey results on how homes are heated. How do those results compare to the graph? What might be the reason for any differences?

Practice

1. What surprises you about the graph from Build Your Understanding, especially in the last 20 years?

2. Look again at the data in Lesson 7. It is from 2000. Compare it to the last year on this graph. Make a chart comparing the two sets of data from 1996 and 2000.

3. Based on your results in question 2, what changes do you predict in energy consumption in the next 5 years? 10 years? 50 years?

4. Explain what strategies you used to come up with your predictions in question 3.

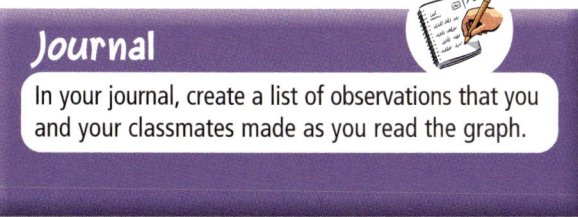

Journal

In your journal, create a list of observations that you and your classmates made as you read the graph.

Lesson 8: Surveys and Graphs

Show What You Know

Review: Lessons 5 to 8, Measurement, Numbers, and Data

1. How much water can the container shown below hold?

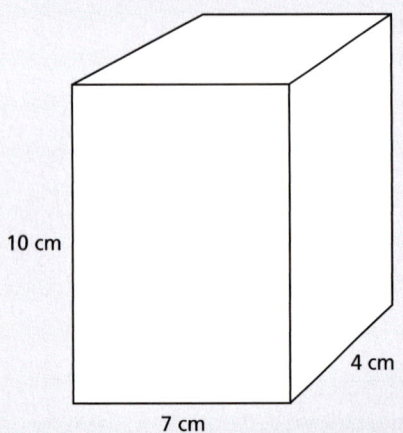

2. a) How many millilitres is 80 cm³? Explain how you know your answer is correct.

   b) What might the dimensions of the container be? Use pictures, numbers, and words.

3. Imagine that a school uses 10 kg of paper at $15.25 per kilogram. If next year the school uses $\frac{1}{2}$ of the paper, how much would it save?

4. Convert the information from your survey in Lesson 8 to fractions.

5. Explain how you converted each number to a fraction.

6. Convert each fraction to a decimal. You may need to use a calculator and round off each number to the nearest tenth.

7. What decisions do you have to make when designing a survey?

8. How can graphs be used to make predictions?

**GEOMETRY AND SPATIAL SENSE
PATTERNING AND ALGEBRA**

Lesson 9
Patterns and Tessellations

ENERGY NOTE

PLAN:
You will create tessellated patterns on window coverings.

DESCRIPTION:
Heat escapes through windows, which is why we use double-paned glass and storm windows. Thick window coverings like curtains can also conserve heat. The people in Mathford enjoy being creative when they make window coverings.

Get Started

1. Look at the window coverings in this illustration. Describe to your class the geometric shapes used.

2. Patterns can be found in fabrics all around you. Look for patterns on your clothes.

3. Where do you see patterns in your classroom?

4. Which of those patterns reflect (flip), translate (slide), or rotate (turn)?

5. Which ones are tessellated patterns? How do you know?

Vocabulary

tessellation: A repeating pattern of closed figures that covers a surface with no gaps or overlaps

Lesson 9: Patterns and Tessellations   **379**

Build Your Understanding

Create Repeating Patterns

You Will Need
- pattern blocks
- coloured pencils

Window coverings conserve energy by keeping heat in during the winter and out in the summer. But it's much more interesting on the inside if you are creative about your coverings!

1. Decide whether you are designing a pattern for a curtain, blinds, or a shade.

2. Each design must completely cover a letter-sized (216 mm x 279 mm) piece of paper.

3. **a)** Use pattern blocks. Start with one pattern block and trace it onto your piece of paper.

   b) Now create a pattern by tracing the block again and again until it covers the entire piece of paper. (Some parts of your pattern may not fit perfectly on the page.) There must not be any gaps or overlapping in your design. The pattern you have created is tessellated.

4. Now use at least two different shapes to create another pattern. Repeat the pattern across the page. You may choose to use more than two shapes.

5. Use a different colour for each type of shape.

Technology

Use a paint/draw computer program to create a curtain design using the draw tools. Copy and paste your shape repeatedly. Use the appropriate tools to rotate, translate, and reflect the pattern piece as needed to create your curtain.

What Did You Learn?

1. In a small group, describe how you constructed your pattern. Use pictures, numbers, and words to explain your pattern.

2. What is the area and perimeter of each pattern block that you used?

3. What does your pattern look like as you turn it 90 degrees? 180 degrees? 270 degrees?

Journal

Look around for patterns that reflect, translate, and rotate. Use pictures to record what you find in your journal.

Practice

Copy and continue each pattern in your notebook. For each pattern, write the pattern rule.

1. 35, 42, 49, ■, ■, ■, ■
2. 25, 36, 49, ■, ■, ■, ■
3. 1.5, 2, 2.5, ■, ■, ■, ■
4. 0.03, 0.06, 0.09, ■, ■, ■, ■

Extension

5. Draw a line or lines on your pattern from Build Your Understanding to show lines of symmetry.

6. Create a collage of geometric patterns. Use magazines that you have permission to cut up.

   a) Is your collage a tessellation? Explain why or why not.
   b) Name and describe all of the geometric shapes that you can find in your collage.

DATA MANAGEMENT AND PROBABILITY

Lesson 10

Analyzing and Graphing More Data

ENERGY NOTE

PLAN:
You will find the mean for a set of data about energy use at different times of the year. You will also graph and analyze the data.

DESCRIPTION:
We use different amounts of energy at different times. Think about what time of year you use the most energy. The people of Mathford know this is useful information.

Get Started

Electric companies keep records on how much energy is used from minute to minute for every day of the year.

1. Review the graph from Chapter 7, Lesson 5 with your class.

2. Describe how you think it was constructed.

3. What data do you need to construct this type of graph?

Journal

Record your predictions about what time of day you think energy use would be highest. Explain your predictions.

Chapter 8: Seedling Sums

Build Your Understanding

Graph and Analyze Data

	Monthly Peak Energy Demand (megawatts)						
	2001	2000	1999		2001	2000	1999
Jan.	22 672	23 428	23 308	Aug.	25 269	23 222	21 225
Feb.	21 935	21 869	21 135	Sept.	19 734	23 191	21 301
March	21 361	20 330	20 615	Oct.	21 286	19 399	18 689
April	18 906	19 307	18 059	Nov.	21 275	22 005	21 160
May	19 163	20 343	19 622	Dec.	22 369	23 291	22 067
June	23 608	21 869	22 846	Mean		21 661	21 122
July	24 013	21 683	23 435	Annual peak		23 428	23 435

This chart shows the data for the monthly peak demand that one electric company recorded for every month from 1999 to 2001. The energy flow is measured in megawatts. Each day, the 20-min peak is recorded—that's the highest megawatt reading for any 20 min in the day. At the end of the month, the peak amount is recorded.

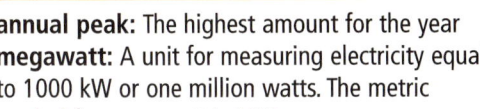

Vocabulary

annual peak: The highest amount for the year
megawatt: A unit for measuring electricity equal to 1000 kW or one million watts. The metric symbol for megawatt is MW.
monthly peak: The highest amount for the month

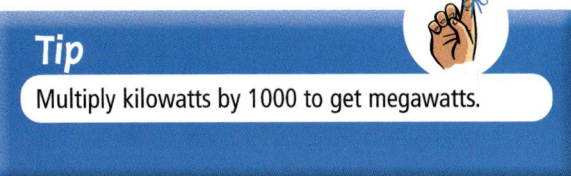

Tip

Multiply kilowatts by 1000 to get megawatts.

1. Find the mean and the annual peak for 2001. Use a calculator to help you.

2. Choose one year from this chart. Graph the data for that year.

Lesson 10: Analyzing and Graphing More Data   383

3. In your notebook, order the monthly energy peaks from maximum to minimum for 2001.

4. Which month had the highest energy peak each year? What may be the cause?

5. Which month had the lowest energy peak each year? What may be the cause?

6. What happened to the energy consumption in April? What might be the reason?

7. Look at the data for the summer months. What pattern do you notice? Which year had the warmest summer? How do you know?

8. What other observations can you make? Describe any data that surprised you. Is there any data that you expected to see? Why?

Journal

In your journal, record what decisions you had to make about how to display the data on the graph.

What Did You Learn?

1. Share with the rest of your class your decisions when choosing how to display the data on the graph.

2. Is it easier to analyze the data in chart form or graph form? Why?

Practice

Extension

1. Graph the data from the chart in this lesson's Build Your Understanding on one graph. Use the graph from Lesson 5 in Chapter 7 as a model to help you.

2. Write three questions that can be answered by looking at your graph. Give them to a classmate to answer.

3. Look at all the graphs you constructed. Predict which month will have the peak energy use in 2002. How does knowing this in advance help the electric company?

DATA MANAGEMENT AND PROBABILITY

Lesson 11

Examining Data From a Table

ENERGY NOTE

PLAN:
You will analyze data about energy peaks from a table.

DESCRIPTION:
You saw how energy use peaks throughout the year. It also peaks on a daily basis. Find out when the daily peaks are in Mathford.

Get Started

Data helps us to keep track of things. Discuss as a class why an energy company would want to keep track of when people use energy the most. How can it help the company plan?

Think of a restaurant. If everybody ate at the restaurant at exactly the same time every day, what would happen? What would the people working at the restaurant have to do to be able to serve everyone at the same time? Share and discuss your answers with the class.

Lesson 11: Examining Data From a Table   385

Build Your Understanding

Analyze Data

You are still working for the Mathford Electric Company. Your job is to analyze the data on this chart and prepare a short report. In your report, you will analyze the pattern of energy use.

	Monthly Peak Energy Demand Time of 20-Min System Peak						
	2001	2000	1999		2001	2000	1999
Jan.	5:40 P.M.	5:40 P.M.	5:35 P.M	July	3:35 P.M.	3:35 P.M.	3:35 P.M.
Feb.	6:45 P.M.	5:50 P.M.	6:45 P.M.	Aug.	3:30 P.M.	2:50 P.M.	3:40 P.M.
March	6:45 P.M.	6:40 P.M.	6:55 P.M.	Sept.	3:40 P.M.	12:40 P.M.	3:30 P.M.
April	7:15 P.M.	7:20 P.M.	7:10 P.M.	Oct.	5:40 P.M.	5:45 P.M.	4:50 P.M.
May	3:45 P.M.	3:20 P.M.	3:35 P.M.	Nov.	5:25 P.M.	5:25 P.M.	5:35 P.M
June	3:17 P.M.	3:40 P.M.	3:35 P.M.	Dec.	5:35 P.M.	5:35 P.M.	5:40 P.M

1. Look at this chart carefully. What time period appears most often? Why? What may be happening during that time period?

2. When does the time period change? Why do you think this may be happening?

3. Share your report with a classmate.

Journal

In your journal, record your observations about energy use in Mathford.

Technology

Using a timeline application, create a timeline of your activities during the day when you are using electricity. Think about how these activities are different in different seasons of the year. Present your timeline to the class, and explain how it might change over the year.

Chapter 8: Seedling Sums

What Did You Learn?

1. What month has the greatest time range across all three years? What month has the least time range? Explain how you figured this out.

2. How can data like this help you conserve energy?

Practice

Write the next three numbers in each set in your notebook.

1. 2, 4, 6, ...
2. 10, 15, 20, ...
3. 48, 56, 64, ...
4. 36, 45, 54, ...
5. 81, 72, 63, ...
6. 22, 20, 18, ...

7. Explain the multiples used in each question above.

8. How are questions 1 (a) and (f) similar and how are they different?

Extension

9. a) Graph the data from the table on the previous page on a multiple line graph. Use a different coloured pencil for each line.

    b) What does a multiple line graph show that is not as clear on a table?

10. Explain how you made your multiple line graph. What decisions did you have to make before you plotted the data? What changes might you make if you were to graph the data again?

Chapter Review

1. When you were hired to plant trees, you were paid $0.25 a seedling. The company now gives its workers a choice of payment: earn $10.00 the first day and then double the amount each day after that *or* earn $0.25 for each seedling that you plant. If you know you can plant 1000 trees in 5 days, which payment plan should you choose? Show your work and write a statement stating which plan you choose.

2. Find the following:
   a) $\frac{1}{4}$ of 12   b) $\frac{1}{4}$ of 24   c) $\frac{1}{4}$ of 48   d) $\frac{1}{4}$ of 32
   e) $\frac{1}{2}$ of 6   f) $\frac{1}{2}$ of 12   g) $\frac{1}{2}$ of 24   h) $\frac{1}{2}$ of 48

3. Find the following:
   a) $\frac{1}{3}$ of 12   b) $\frac{2}{3}$ of 24   c) $\frac{1}{3}$ of 30   d) $\frac{2}{3}$ of 48

4. Answer the following questions:
   a) 7946 ÷ 8   b) $0.72 x 5   c) $3.15 x 4   d) 964 ÷ 4

5. a) Draw the reflection, rotation, and translation of the shape on the right, and label each transformation. Be as detailed as you can.

   b) How many lines of symmetry does the shape on the right have? Draw a similar shape and indicate the lines of symmetry.

6. ○ = $\frac{1}{2}$ △ = $\frac{1}{3}$ □ = $\frac{1}{4}$

Use these shapes to draw a picture of a house that has a value between 4 and 6. Calculate the total value of the house that you have made.

7. If it takes 5 s for 1 L of water to come out of a hose, how much water would you use in half an hour? Use pictures, numbers, and words.

Chapter 8: Seedling Sums

8. Convert the following measurements:
   a) 200 mm = ■ m   b) 75 cm³ = ■ mL
   c) 9.4 m = ■ mm   d) 34 000 m = ■ km

9. Write the following as fractions:
   a) 0.10   b) 0.24   c) 0.16   d) 0.35   e) 0.56

10. Write each as a mixed number:
    a) 7.6   b) 154.8   c) 345.89   d) 209.90   e) 870.04

11. Write the following in words:
    a) 26.4   b) 408.7   c) 1890.6
    d) 89.45   e) 567.09   f) 5348.20

12. The figures in the chart tell us how much of the earth's water supply comes from different types of water.
    a) Order these numbers from greatest to least.
    b) Round off each number in the chart to the nearest tenth of a percent.

Earth's Water Supply	
Type of Water	Percentage
oceans	97.29%
groundwater	0.51%
lakes and streams	0.01%
glaciers and ice caps	2.19%

13. Create a tessellated pattern using pattern blocks. Using math terms, describe your pattern and explain how you know it is a tessellation.

14. a) Graph the following data on the depth of the Great Lakes.
    Lake Huron: 229 m   Lake Superior: 405 m
    Lake Michigan: 281 m   Lake Erie: 64 m
    Lake Ontario: 244 m

   b) Calculate the mean depth.

Chapter Review

Chapter Wrap-Up

You have reached the end of Chapter 8. In this chapter, you have learned a lot about how much energy we use and how we can use less energy. You have read data tables and used multiplication.

You are still employed by Mathford Electric Company, but you have received a promotion. Since you have shown excellent mathematical skill in creating, analyzing, and evaluating graphs, you are now in charge of the research department.

The data you must work with lists the hourly energy consumption for Mathford. You've just been handed the data for January 1, 2, and 3, 2003 in the chart below.

Times	Jan. 1, 2003 megawatts	Jan. 2, 2003 megawatts	Jan. 3, 2003 megawatts	Times	Jan. 1, 2003 megawatts	Jan. 2, 2003 megawatts	Jan. 3, 2003 megawatts
12:00 A.M.	16 434	16 327	17 267	12:00 P.M.	16 962	20 478	20 401
1:00 A.M.	15 881	15 975	16 689	1:00 P.M.	16 887	20 376	20 525
2:00 A.M.	15 349	15 843	16 396	2:00 P.M.	16 824	20 237	20 476
3:00 A.M.	15 060	15 842	16 207	3:00 P.M.	16 971	20 286	20 481
4:00 A.M.	14 893	15 936	16 283	4:00 P.M.	17 884	20 925	21 136
5:00 A.M.	14 955	16 582	16 759	5:00 P.M.	19 459	22 219	22 364
6:00 A.M.	15 189	17 741	17 861	6:00 P.M.	19 624	22 197	22 352
7:00 A.M.	15 506	19 169	19 149	7:00 P.M.	19 478	21 764	21 938
8:00 A.M.	15 691	19 959	19 734	8:00 P.M.	19 252	21 271	21 524
9:00 A.M.	16 176	20 268	20 128	9:00 P.M.	18 795	20 454	20 681
10:00 A.M.	16 577	20 507	20 268	10:00 P.M.	18 096	19 336	19 517
11:00 A.M.	16 870	20 560	20 382	11:00 P.M.	17 056	18 345	18 365

Chapter 8: Seedling Sums

Your task is to analyze and evaluate the data. You must also create a multiple-bar graph to present the data to the Mathford town council.

Use charts to show the rest of the data that you collect from this table. Work in a small group of three students.

1. Find the total amount of energy used each day.
2. Find the mean amount of energy used each day.
3. Find the hour with maximum usage for each day.
4. Find the hour with minimum usage for each day.
5. Graph each day's usage.
6. Make a multiple-bar graph for the hours between 9:00 A.M. and 12:00 P.M. for the three days.
7. Write several observations you can make from this data. When is the most electricity being used? What happens throughout the day? When does consumption start to increase and decrease? Why? When is consumption at its lowest? Why?
8. Write three suggestions for lowering energy consumption.
9. Create a poster, make a song, or write a play to present your suggestions for lowering energy consumption.
10. Think back to all the work you have done in Chapter 8. What has been the most helpful for you in this project?

Technology

Using a spreadsheet computer program, input the data from steps 1 to 4 to create the graphs in steps 5 and 6. Present your findings in a multimedia slide-show presentation.

Chapter Wrap-Up

Chapter 9

Energy by the Numbers

Conserving energy means using our resources wisely. Renewable resources, such as water and trees, and non-renewable resources, such as oil and gas, are in good supply in Mathford and in most parts of Canada.

In this chapter, you will
- predict and explore probability
- analyze, measure, and compare data using multiplication
- analyze, examine, and calculate chart data
- examine and analyze patterns

At the end of this chapter, you will brainstorm a list of what you learned in this unit about energy. You will then prepare a report that outlines ways to lower energy consumption and costs.

1. In your journal, explain probability, and write about an example.
2. Work with a classmate and create a pattern on a piece of paper. What makes your design a pattern?

Chapter 9: Energy by the Numbers

DATA MANAGEMENT AND PROBABILITY

Lesson 1
Predicting Probability

ENERGY NOTE

PLAN:
You will predict probability.

DESCRIPTION:
Canada has many natural resources that supply us with energy.

The amount of energy we have available is affected by where we live. For example, there might be greater probability (or chance) of having more energy available to us if we live in a certain area.

Get Started

You Will Need
- paper clip
- paper
- pencil
- reproducible sheet with spinners

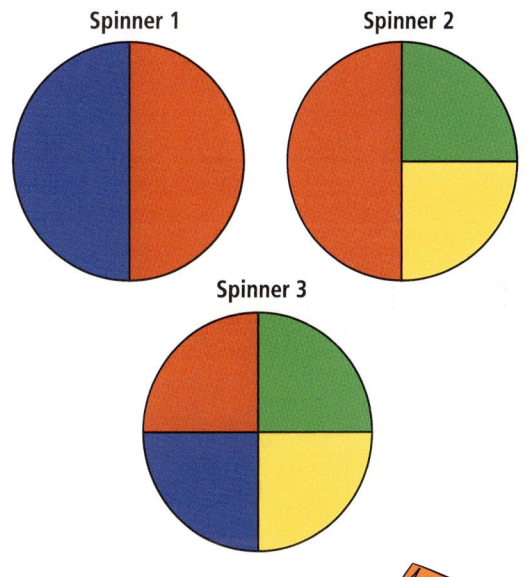

Look at these spinners one at a time. What is the probability of spinning blue in Spinner 1? Spinner 2? Spinner 3? Explain your answer for each.

With a partner, use the reproducible sheet to make three spinners like these and check your predictions. To spin your spinners, put a paper clip in the centre of the spinner and hold it in place with a pencil. Then, spin the paper clip. How can you keep track of your spins? How many spins do you have to make before you can accurately predict probability? Share and discuss your answers with the class.

Vocabulary

probability: A number statement that tells how likely it is that an event will take place
probability statement: A way of saying your predictions. If you think there's a 1 in 5 chance of an event happening, you say there's a 1 in 5 probability.
sample: Part of the whole group. When determining probability, the larger the sample, the more accurate the results.

Lesson 1: Predicting Probability

Build Your Understanding

Predict Probability

You Will Need
- paper
- paper clip
- pencil
- reproducible sheet with spinner

People use different sources of energy to heat their homes. The type of energy you use for heating is your choice. But what if it were left up to chance or probability?

1. Make a spinner like the one shown below, using the reproducible sheet, a paper clip, and a pencil.

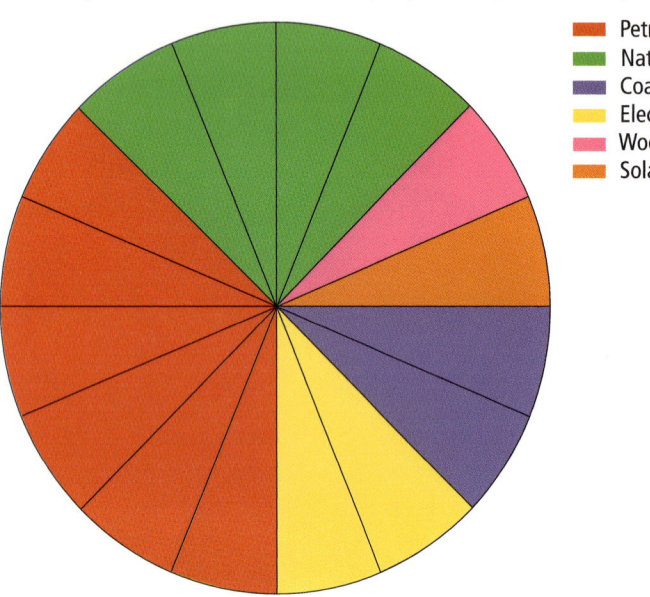

- Petroleum (Oil)
- Natural Gas
- Coal
- Electricity
- Wood
- Solar

2. Spin the spinner 20 times. Keep track of your spins. When you are done, tally your results.

3. After you have completed 20 spins, write a probability statement and fractions for your results. Your data is called a sample. It is called a sample, because the data is not how many times you could spin (you could keep spinning forever!), but it is the number of times you did spin to get results.

394 Chapter 9: Energy by the Numbers

4. Spin 20 more times. What were the results this time? Did the sample change? How?

5. Now combine your data with your classmates' to make a larger sample. Collect the data from the spins your classmates did. How does a larger sample affect the probability statement?

6. Express your results in fractions.

Tip

If your data shows that after 20 spins, you land on coal once, you have a 1 in 20 slight probability of heating your home with coal, or $\frac{1}{20}$ or 1:20. A $\frac{1}{20}$ probability would be written as the ratio 1:20.

What Did You Learn?

1. Discuss the results of this activity with your class.
2. Explain how you made a sample.
3. Explain how you expressed the results in fractions.
4. Summarize your results in a paragraph, and share it with a classmate.

Practice

1. **a)** If you put your hand into a bag with only 2 marbles, 1 red and 1 black, what is the probability that you will choose a green marble?
   b) What is the probability you will choose a red one?
   c) A black one?

2. You have a bag of 20 assorted tulip bulbs. There are 8 red tulips, 3 white tulips, 2 purple tulips, and the rest are orange. You select 10 bulbs to plant in the fall, but you don't know which colours you've chosen. When they bloom in the spring, what is the probability that you will have 4 red tulips? Explain how you got your answer.

Technology

Using a computer application of your choice, create a tally chart with the headings Sum, Tally, and Frequency. Work with a classmate, and roll a set of two number cubes 72 times. Record your results on the chart.

Lesson 2

Using Multiplication to Analyze Data

ENERGY NOTE

PLAN:
You will analyze data about water using multiplication.

DESCRIPTION:
A lot of water is available in Canada. Our country is surrounded on three sides by ocean, and we have some of the largest lakes in the world. In many parts of Canada, water is used to produce electricity. But water has many other uses, as you will see in Mathford.

Did You Know?

- The average family does 392 loads of laundry in a year.
- Each of us flushes 175 L of water a day down the drain.
- A single lawn sprinkler spraying 19 L per minute uses 50 percent more water in just one hour than a combination of 10 toilet flushes, two 5-min showers, two dishwasher loads, and a full load of laundry.
- It takes 295 000 L of water to produce 130 kg of paper.
- A leaky tap drips one drop of water every 6 s. One drop has a volume of 0.3 mL.

Get Started

Amount of Water Used per Day per Person	
United States	382 L
Canada	343 L
Italy	250 L
Sweden	200 L
France	150 L
Israel	135 L

1. Look at the table to the right. Discuss the data with your classmates.

2. How much water would you use in one year? Explain how you calculated your answer.

3. Graph the data. How does graphing help you understand the data?

Chapter 9: Energy by the Numbers

Build Your Understanding

Analyze Data

Use a calculator and the data in Get Started and Did You Know? to answer these questions. Show your work in pictures, numbers, and words.

1. How much water would your family use in one year?

2. How much would the families of all your classmates use in one year? Explain how you calculated your answer.

3. If everyone in your family used 5 L of water less each day, how much water would your family save in one year?

4. If you do your laundry at a laundromat, imagine that you pay $1.25 to wash a load and $1.50 to dry a load. How much would it cost to wash and dry 392 loads of laundry?

5. The average North American uses more than 260 kg of paper a year. How much water is used to produce paper for one person? How much water is used to produce enough paper for your class?

6. If your family moved to France for one year, how much water might your family use?

What Did You Learn?

1. Check your answers with a partner. Recalculate if you disagree so that you can reach a consensus.

2. What strategies did you use to find the answers to the Build Your Understanding questions?

3. How can estimating help you check your work?

4. What data surprised you? Why?

Practice

1. If you take 25 showers a month, how old will you be when you have taken 1000 showers? Use pictures, numbers, and words.

Tip
1 mL = 1 cm³
1 L = 1000 cm³

Based on the information given on page 396, calculate the number of millilitres of water that drip from a leaky tap in

2. one minute
3. one hour
4. one day
5. one year
6. Convert your answers for questions 2 to 5 to cubic centimetres.

Write eight equations linking the three numbers in each set—four addition questions and four subtraction questions. The first one is done for you.

7. 0.2, 0.4, 0.6

   Answers:
   0.2 + 0.4 = 0.6   0.6 − 0.4 = 0.2
   0.4 + 0.2 = 0.6   0.6 − 0.2 = 0.4
   0.6 = 0.2 + 0.4   0.2 = 0.6 − 0.4
   0.6 = 0.4 + 0.2   0.4 = 0.6 − 0.2

8. 7, 8, 15
9. 0.54, 0.38, 0.92
10. 8, 4, 12
11. 1, 0.75, 0.25

Chapter 9: Energy by the Numbers

DATA MANAGEMENT AND PROBABILITY
NUMBER SENSE AND NUMERATION

Lesson 3
Working With Water Data

ENERGY NOTE

PLAN:
You will examine a circle graph and use multiplication to calculate water use.

DESCRIPTION:
Find out how much water is needed for a daily task in your town or in Mathford.

Get Started

In a small group, examine the circle graph. Discuss why the results might be as they are. Think about how you use water. How accurately do these figures represent your own patterns of water use?

Journal
Write two statements about your water use. Share your statements with other classmates.

Tip
The circle in a circle graph represents all of something, and each section shows a part of the whole. The parts can be measured as fractions.

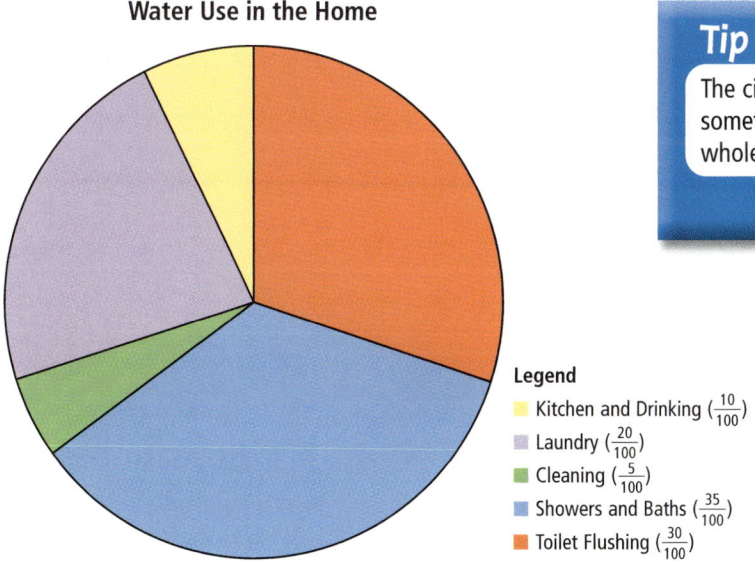

Water Use in the Home

Legend
- Kitchen and Drinking ($\frac{10}{100}$)
- Laundry ($\frac{20}{100}$)
- Cleaning ($\frac{5}{100}$)
- Showers and Baths ($\frac{35}{100}$)
- Toilet Flushing ($\frac{30}{100}$)

Lesson 3: Working With Water Data

Build Your Understanding

Multiply Water

A family of four can use a lot of water. Find out how much one family uses and how much it costs.

With a partner, copy and complete the chart below. Calculate how many litres are used in one day, one week, and one year for each task. You can use a calculator to help you.

Type of Water Use	Average Amount for One Use	Number of Times Used in a Day	Total Number of Litres Used in a Day	Total Number of Litres Used in a Week	Total Number of Litres Used in a Year
toilet flush	17 L	16			
shower	75 L	3			
tub bath	112 L	1			
hand washing	8 L	30			
teeth brushing	1 L	12			
dishwasher	40 L	1			
dishwashing by hand	30 L	1			
cooking a meal	20 L	3			
getting a drink	1 L	15			
cleaning house	30 L	1			
washing machine	120 L	1			
Total					

What Did You Learn?

1. How much water does a family of four use in one year?
2. What statements can you make about our water usage? Share your statements with another pair of students.

Chapter 9: Energy by the Numbers

Practice

Extension

1. Compare your results from Build Your Understanding with the data in Lesson 2. Do they match? What might account for the difference?

2. If a lawn sprinkler uses 19 L of water a minute, and an average family waters their lawn for 1 h twice a week for 10 weeks of the year, how much water is used in a year? Use a calculator to help you. Show your work in pictures, numbers, and words.

3. Alexis uses 150 L of water a week to shower. How much water does she use in half a year? Show your work in pictures, numbers, and words.

4. How might you calculate the average amount of water your class uses a day at school? Use the data from Lessons 2 and 3 to help you. Share your strategies with the class.

5. Use your strategy or a classmate's strategy to calculate the average amount of water your class uses per day at school. Share and compare your answers with a partner.

Lesson 4

Comparing Data Using Multiplication

ENERGY NOTE

PLAN:
You will use multiplication to compare water use.

DESCRIPTION:
Find out how much water it takes to wash some dishes, either using a dishwasher or washing the dishes by hand.

Get Started

Read the facts in the Did You Know? box out loud with a classmate.

Did You Know?

- Washing dishes by hand uses 30 L of water.
- A dishwasher uses 40 L of water.
- A dishwasher saves an average of four hours of hand washing per week.

Build Your Understanding

Calculate Dishwater

Look at the facts in the Did You Know? box.

1. What are the advantages and disadvantages of using a dishwasher?

2. Compare the amount of water used washing dishes by hand to the amount used with a dishwasher. If you choose to wash your dishes by hand, how much water can you save?

Technology

Using the Internet and other sources, gather information on three or four brands of dishwashers to find the best buy for your family. Remember that not only is price important, but also how much water a brand uses, since some people have to pay for water as well. Record your results on a spreadsheet.

3. If you wash dishes twice a day, how much water would you save in a week?

4. How much would you save in a year?

5. If you used a dishwasher, how many hours could you save in a year? How many days is that?

What Did You Learn?

1. Check your answers with a partner. Recalculate if you disagree so that you can reach a consensus.

2. Explain the difference in water usage and time between washing dishes by hand and washing dishes in the dishwasher, using pictures, numbers, and words.

Practice

Use pictures, numbers, and words to answer the questions below.

1. Perhaps you could save the environment and also make a profit. Imagine that you offer to do the dishes. You tell your parents that you'll charge them $0.02 for the first day. Each day you'll charge twice as much as the day before. How much would you have earned in 2 weeks? in a month? What is happening to your earnings?

2. When you offered to do the dishes, your parents thought they had a great deal for just pennies a day. Explain why you may soon be out of a job.

Lesson 4: Comparing Data Using Multiplication   403

Show What You Know

Review: Lessons 1 to 4, Probability and Multiplying Data

1. Design your own spinner to find the probability of something you are interested in. Graph your results.

2. **a)** Imagine that your class washed 78 cars and trucks for a fundraiser. You used about 110 L of water for each car and 150 L for each truck. There were twice as many cars as trucks. How many cars and trucks did the class wash?

   b) How many wheels did the class wash?

3. Calculate the amount of water used washing the cars and trucks.

4. If the donations were $4.00, $6.00, and $8.00, what was the mean donation?

Lesson 5
Problem Solving With Patterning

ENERGY NOTE

PLAN:
You will use your problem-solving skills to calculate rainfall and examine patterns.

DESCRIPTION:
Look at annual rainfall for some communities in Canada — and in Mathford.

Did you know that the wettest place in Canada is Ocean Falls, British Columbia, which has an average of 4386.8 mm of precipitation per year? The place with the least amount of precipitation is Eureka, Northwest Territories, with an average of 64 mm per year.

Get Started

You Will Need
- calendar

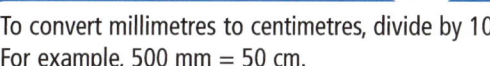

Tip
To convert millimetres to centimetres, divide by 10. For example, 500 mm = 50 cm.

Some communities don't need to worry about water consumption because they have too much water! Work in small groups to calculate these problems. You may use a calculator to help you.

1. What is the mean amount of rainfall per day for each community listed in the Energy Note?

2. If it rained every day in Ocean Falls, and an equal amount of rain fell each day, about how much rain would fall in January?

3. If it rained only once every 10 days all year long, how many days of rain would each community have? About how much rain would fall in Ocean Falls on the days that it rained? in Eureka?

4. How many centimetres of rain, on average, fell in one year in each community?

Build Your Understanding

Record Rain in Mathford

Weather can be difficult to predict, but the residents of Mathford have begun to notice something strange about their weather. It has rained every fourth day from Tuesday, April 1 to Saturday, June 28.

Use a calendar to help you answer the following questions:

1. How many days did it rain in Mathford?

2. On what day of the week did residents receive their fourth day of rain? seventh day of rain? eleventh day of rain?

3. Did it rain on May 1 or June 1?

4. What patterns did you find as you solved these problems?

5. If 9 mm of rain fell during the first rainstorm, and 11 mm during the next storm, and the amount of rain that fell during each rainstorm after that increased by the same amount, how much rain fell during April, May, and June?

6. Convert your results from question 5 to centimetres.

7. How much rain in total had fallen after the fourth day of rain? seventh day of rain? eleventh day of rain?

8. Wednesday, April 2 and every third day after was partly sunny and partly cloudy. How many partly sunny and partly cloudy days did Mathford have from April 2 to June 30?

9. How many days were both partly sunny, partly cloudy and rainy?

What Did You Learn?

1. Share with a small group of three or four classmates how you solved the problems. What approach helped you in solving the problems?
2. When did you use a calculator? When was it faster to calculate your answers without a calculator?

Journal

Explain how you solved questions 8 and 9 of Build Your Understanding. What strategies did you use?

Practice

Copy and convert the following in your notebook:

1. 1 m = ■ mm
2. 5000 mm = ■ cm
3. 25 cm = ■ m
4. ■ mm = 9.5 m
5. ■ km = 16 000 m
6. 850 m = ■ km
7. 0.05 m = ■ mm
8. ■ m = 10.34 km

Write eight equations linking the three numbers in each set—four multiplication questions and four division questions.

9. 2, 3, 6
10. 6, 7, 42
11. 64, 89, 5696
12. 30, 80, 2400

Extension

13. a) Find out about how much rain falls in your city or community in a year.
    b) How much falls in your province?
    c) How has the amount of precipitation changed over the years?
    d) Where could you find this information?

14. a) Find out about how much rain falls in a different city or community during one year.
    b) Compare these results to those in your city or community.
    c) What conclusions can you make based on your findings?

PATTERNING AND ALGEBRA

Lesson 6

Finding Patterns

ENERGY NOTE

PLAN:
You will find patterns and use multiplication to calculate umbrella sales.

DESCRIPTION:
It's still raining in Mathford, and the general store is selling lots of umbrellas.

Get Started

You Will Need
- calculator
- chart paper
- markers

Storeowners make money according to how much they sell. They hope that they have stocked their stores with the kinds of items that people need or want to buy.

For example, school lunch items are popular items in grocery stores. Let's say that you brought a new type of snack bar to school. It tastes great and you shared one with your friend. The next day he or she brought the new type of snack bar to school in his or her lunch as well. Each of you then shared your bars with another friend. The day after that, four students had the new type of snack bars in their lunches.

Chapter 9: Energy by the Numbers

1. What is the pattern?

2. If the pattern continued, how long would it be before your 32 classmates had the new type of snack bars in their lunches? How long before the entire school of 256 students were bringing snack bars?

3. Work with a partner to solve this problem. What problem-solving strategies will you use? Use chart paper to do your calculations. Share your answers with the class.

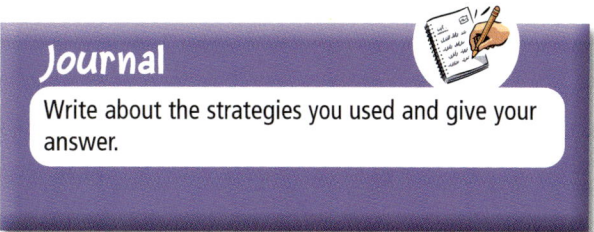

Journal

Write about the strategies you used and give your answer.

Build Your Understanding

Count Umbrellas

It has been the rainiest April that anyone in Mathford can remember. While the farmers are thrilled to see so much water for their fields and crops, the people of Mathford are trying to stay dry.

The illustration below shows how many umbrellas were sold by the Mathford General Store on four rainy days in Mathford.

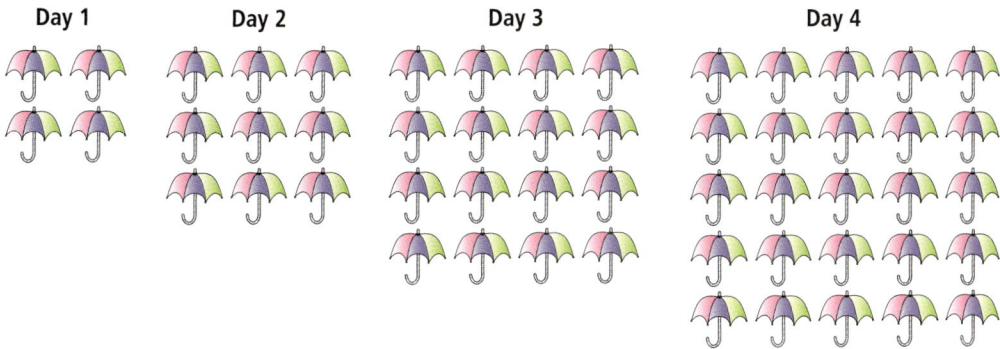

Use pictures, numbers, and words to answer these questions.

1. What is the pattern?

2. If the pattern continues, how many umbrellas will the store sell on the last day of April?

3. How many umbrellas will the store have sold by the tenth day? the twentieth day?

Lesson 6: Finding Patterns

4. How many umbrellas will the store sell during the entire month of April?

5. If each umbrella sells for $3.89, how much money will the owner make in umbrella sales in April?

6. If the owner paid her supplier $1.29 for each umbrella, how much will she have made?

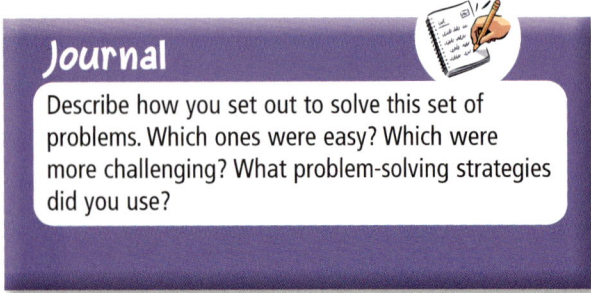

Journal

Describe how you set out to solve this set of problems. Which ones were easy? Which were more challenging? What problem-solving strategies did you use?

What Did You Learn?

Explain your problem-solving process to a classmate using the data you've collected.

Practice

1. If you made a graph of the umbrella sales, what type of graph would you make? Why? Describe what it would look like.

2. Would the sales of all items sold in a store have a similar graph? Why or why not?

3. What factors affect sales of products?

4. Find the daily mean (average) of umbrellas sold in April.

5. Calculate the 7% GST and the 8% PST (provincial sales tax) for one umbrella.

6. Calculate the sales taxes for all the umbrellas sold.

Lesson 7
Probability Problems

DATA MANAGEMENT AND PROBABILITY

ENERGY NOTE

PLAN:
You will examine patterns in umbrellas. You will also look at probability to solve problems.

DESCRIPTION:
Lots of rain is filling up the water reservoir in Mathford, and the residents still need their umbrellas as they walk around town.

Get Started

The Mathford General Store sells two types of umbrellas: red ones and green ones. They're sold in sets of two, but they are in a cardboard box, so you can't see what colours you are buying. What is the probability that you will get both a green and a red umbrella? Work out this problem with your classmates.

Tip

You can use a tree diagram to help solve this problem.

Vocabulary

tree diagram: A branching graph showing the possible outcomes in a probability experiment

Lesson 7: Probability Problems 411

Build Your Understanding

Design Patterns

Mathford General Store has just received a new shipment of umbrellas. Customers now have several choices of umbrellas. There are 4 colour choices: red, blue, green, and yellow. Each umbrella has either 2 or 3 colours. On each umbrella, each colour is used equally. Each umbrella has 6 sections.

Work with a partner. Find out how many choices the customer has for umbrellas.

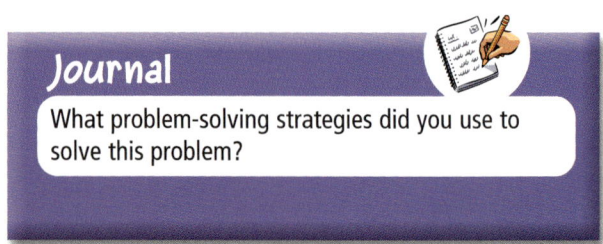

Journal
What problem-solving strategies did you use to solve this problem?

What Did You Learn?

1. There are several ways to solve this problem. With your partner, share the strategies you wrote in your journal.
2. Share your answers with the rest of your class.

Practice

Extension

1. The umbrella factory has been asked to design a 4-colour umbrella like their 2- and 3-colour designs. How many choices will the customer have? Work with a partner.

2. Work as a class to determine how many choices the customer would have if there were 6 colours. Put all your problem-solving strategies together to solve this problem.

Tip
You can draw pictures or make a tree diagram. Use a chart to organize your results.

Chapter 9: Energy by the Numbers

Show What You Know

Review: Lessons 5 to 7, Patterns, Money, and Probability

Use pictures, numbers, and words to answer the questions below.

1. Imagine that it rains 2 mm on April 1 and it rains every second day until the end of the month. By the end of the month, it has rained 48 mm.

   a) What is the pattern in rainfall for the month of April?

   b) How much rain fell in April?

2. Imagine that you just finished designing an umbrella, and now you want to sell it. If it cost you $6.45 to design your umbrella, and you sold your umbrella for $12.99, how much money would you make if you sold 300 umbrellas?

3. The Mathford General Store decided to sell umbrellas in packs of 4, due to all the rain they have been getting. There are 4 new colours of umbrellas: purple, orange, brown, and pink. Each pack is sold in a cardboard box, so you can't see what colours you are buying. What is the probability that you will get one of each colour of umbrella in a pack?

DATA MANAGEMENT AND PROBABILITY

Lesson 8

Ranking and Graphing Data

ENERGY NOTE

PLAN:
You will analyze chart data and find the mean (average) use of groundwater in Canada.

DESCRIPTION:
Lots of rain means lots of groundwater, which is available when the rain soaks into the soil. Lots of groundwater is good for communities like Mathford because they then have plenty of water for their needs.

Get Started

In Canada, 7.9 million people, or 0.26 of the population, depend on groundwater for their water. Almost $\frac{2}{3}$ of them, or five million people, live in rural areas.

1. Look at the chart below to see what it tells about our use of groundwater.

Province/Territory	Population Dependent on Groundwater	Province/Territory	Population Dependent on Groundwater
British Columbia	0.22	Nova Scotia	0.50
Alberta	0.27	Prince Edward Island	1.00
Saskatchewan	0.45	Newfoundland and Labrador	0.29
Manitoba	0.24	Yukon	0.63
Ontario	0.23	Northwest Territories and Nunavut	0.01
Québec	0.22		
New Brunswick	0.64		

Chapter 9: Energy by the Numbers

2. Discuss your observations as a class.

3. Rank the provinces and territories in order from least use of groundwater to greatest use.

Journal

Write down your observations in your journal. Make comparisons between your province or territory and other parts of Canada.

Build Your Understanding

Analyze the Map

You Will Need
- grid paper

Data is portrayed in many different ways. This map of Canada shows the amount of groundwater use for each province and territory.

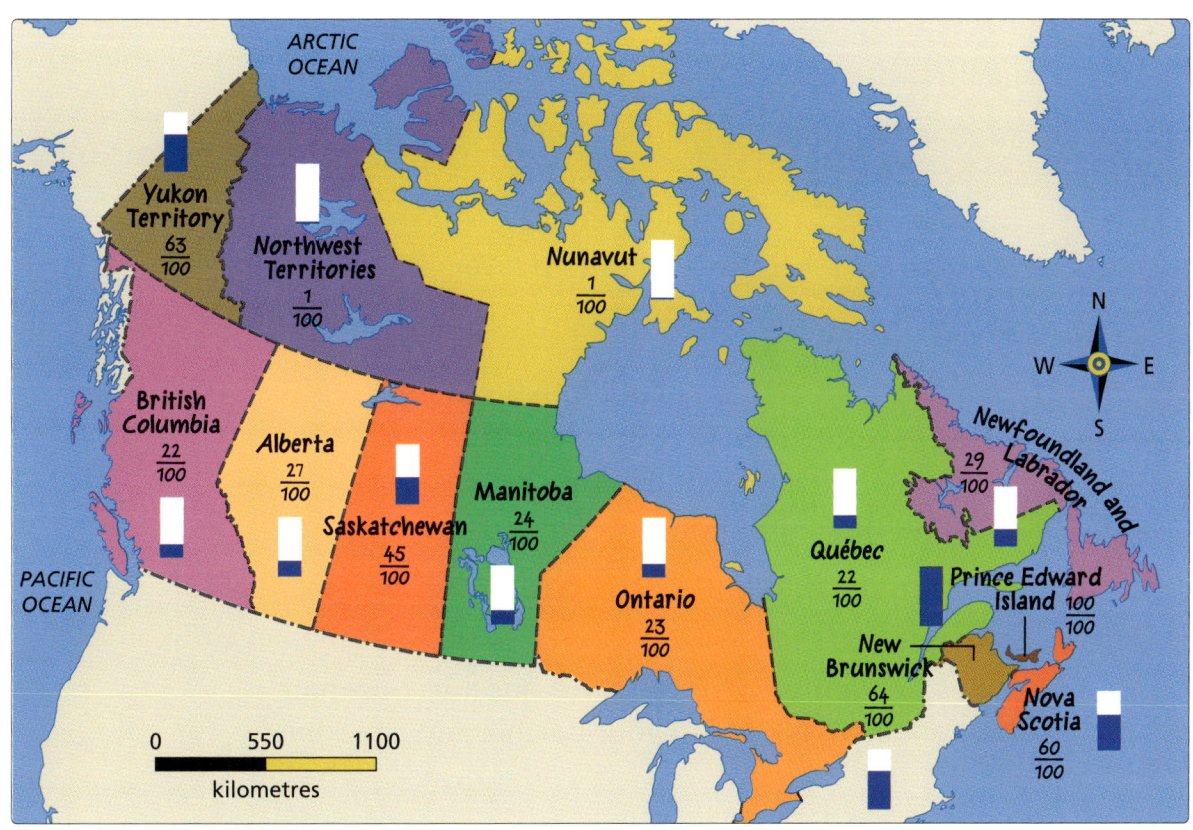

Lesson 8: Ranking and Graphing Data 415

Examine the map and then complete the following:

1. Find the mean groundwater use for all provinces and territories.
2. Write each fraction as a decimal.
3. What fraction of the Canadian population relies on water from sources other than groundwater?
4. For the following groups, find the mean use of groundwater for each group:
   a) Territories
   b) Western provinces
   c) Central Canada (Ontario and Québec)
   d) Atlantic provinces
5. Rank and graph the data from question 4. Which part of Canada relies on groundwater the most? Which part relies on it the least?

What Did You Learn?

1. Check your answers with a partner. Recalculate if you disagree and can't come to a consensus.
2. How did organizing the data help you to understand it?
3. Display the information from the map in another type of graph.

Practice

1. Make one line graph showing:
   a) the fraction of the Canadian population that relies on groundwater
   b) the fraction of the Canadian population that relies on water from sources other than groundwater
2. What are the similarities, differences, and patterns that you notice?
3. Create three questions about your graph for another classmate to answer. Make an answer sheet so your partner can check his or her answers.

Lesson 9

Analyzing and Comparing Data

DATA MANAGEMENT AND PROBABILITY

ENERGY NOTE

PLAN:
You will analyze and compare data about gas prices from a chart.

DESCRIPTION:
The price of gasoline can vary a great deal from one part of the world to another. In Canada, we have large natural gas and oil resources, so gasoline prices are low compared to many other places. Communities like Mathford, which are located quite far from other towns, rely on cars to get around.

Get Started

Data is kept on gas prices around the world. Most Canadians prefer to drive or be driven, rather than to walk or to take the bus. How much does this cost us? We can ask questions to analyze the table shown on the next page.

1. How often has the average price of gasoline gone down from one year to the next? In which countries did this happen?

2. With a classmate, discuss what the data tells you. Now think of three questions about the table. Record the questions in your notebooks.

3. Give your questions to another pair of students to answer.

Lesson 9: Analyzing and Comparing Data   417

International Prices for Motor Gasoline in Cents per Litre (Taxes Included)						
Period	United States	United Kingdom	France	Italy	Germany	Canada
1985	43.30	76.30	87.20	95.40	66.40	53.50
1986	34.10	76.10	95.90	120.10	69.20	47.70
1987	33.30	82.70	106.30	133.20	76.60	50.10
1988	30.90	80.80	99.80	128.30	70.70	49.30
1989	32.10	79.10	95.20	118.20	78.10	51.60
1990	35.60	91.80	112.50	143.20	91.50	58.40
1991	34.60	98.20	110.00	142.10	97.80	57.00
1992	36.10	107.80	120.90	150.70	112.60	54.70
1993	37.80	104.30	124.00	132.20	105.00	53.60
1994	39.90	118.50	138.10	143.70	126.80	52.80
1995	41.70	128.80	160.70	154.40	143.30	55.40
1996	45.00	129.00	165.80	166.80	142.00	58.10

Build Your Understanding

Analyze Data

Work with others in a small group. Use the data presented in the above table.

1. How many years of data does this table present?
2. Find the difference between the highest and lowest gasoline prices for each year.
3. Order the numbers in the Canadian column from least to greatest.
4. Take the figures for Italy and order the decimals from greatest to least.
5. Find out by how much gasoline went up in each country for the years 1994 to 1996. Write the data on a chart.

6. What do you notice about price changes for each country from 1985 to 1996? By how much has the price changed in each country over these years?

7. Find the mean price of gasoline for each country.

8. Write any 10 decimal numbers for France or Italy as mixed numbers and in words.

9. If an average family used 85 L of gasoline a week, how much did gas cost a family in each country for one week in 1996? How much for one month? How much for all of 1996? Show your work.

10. Imagine that the same family switched to a more fuel-efficient vehicle and used only 64 L of gasoline per week. How much money would it save for the year of 1996? Calculate the savings for each country. Show your work.

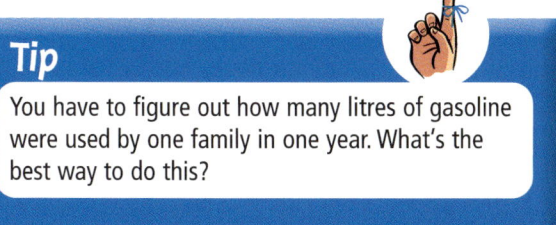

Tip

You have to figure out how many litres of gasoline were used by one family in one year. What's the best way to do this?

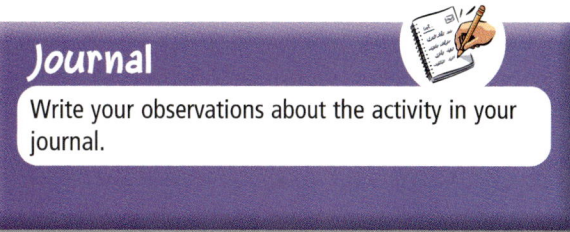

Journal

Write your observations about the activity in your journal.

What Did You Learn?

1. Look at the gasoline prices for 1994, 1995, and 1996 for all countries. What happened?
2. What happened between 1987 and 1988 that did not occur between any other two consecutive years?
3. Which country consistently had the lowest gasoline price?
4. What other observations can you make?

Practice

Extension

1. Find out the price of gas in your neighbourhood. If someone in your family drives a car, keep track of how much gas your family uses in one week. Calculate how much this costs.
2. As a class, gather this data from everyone in the class. How much gasoline did all the families of the entire class use? How much did that cost?
3. Using your data from question 2 as an average, how much gasoline would be used in one year? How much would it cost?
4. Make recommendations as to how you could lower your gasoline costs over the year.
5. Make a double bar graph showing Canada and one other country from the years 1985 to 1996. What do you observe?
6. Some people say gas prices have increased a lot. Have they? Why or why not? How does a chart like this help you make accurate observations?

Technology

Access the Statistics Canada Web site (www.statcan.ca). Use a spreadsheet computer program to record how many litres of gasoline each of the provinces and territories bought last year. Sort the order from greatest amount to least.

Lesson 10
Transportation Survey and Probability

ENERGY NOTE

PLAN:
You will survey a population about transportation and write probability statements.

DESCRIPTION:
One way to conserve gas and save money is to find another way to get around. Mathford is small enough to bike and walk around. In other communities, you can take the bus.

Get Started

Data is gathered from either a sample or from a population. What is the difference? How is the analysis of data affected if you use a sample? Discuss this with a partner.

You will gather data about how you and your family travel. How do you get to all the places you need to go?

Vocabulary

population: The total number of people in a specific area
sample: A part of a population selected to represent the population as a whole

1. Make a list of all your family members and record how they get to work or school. You can present your data using a table like the one shown below. Add additional columns for more family members if needed.

2. What do you notice about the data?

	Family Member 1	Family Member 2	Family Member 3	Me
Public Transit				
Motor Vehicle as Driver				
Motor Vehicle as Passenger				
School Bus				
Bicycle/Scooter/ Skateboard				
Walk				

Build Your Understanding

Survey a Population

1. Sort the data from all the tables of your classmates into one class table.

2. How many people are represented in the data?

3. Rank your data from most common to least common method of travel.

4. Write each statistic as a fraction and a ratio statement.

5. What observations and statements can you make about methods of travel of the families of your class?

Tip

For example, if 50 people use public transit, and a total of 112 people were surveyed, write the number as
a fraction: $\frac{50}{112}$ a ratio: 50:112

422   Chapter 9: Energy by the Numbers

What Did You Learn?

1. Explain how you organized the information from the class table.

2. Why do you think data is sometimes gathered from a sample, rather than from a population?

3. Do you think the data for your class would be similar to data collected in a different classroom? a different school? Explain your answer.

Practice

1. Predict the results for the entire school. What about your entire community? You are now turning your population into a sample. What are the benefits of collecting population data? What are the difficulties?

2. Write probability statements for your data. For example, 50 out of 112 people in the entire city probably use public transit. Write your probability statements as fractions.

Extension

3. If your family has a motor vehicle, keep track of your family's motor-vehicle use for a week. Record how many trips were made, how far each trip was, and how many people were in the vehicle. What do you notice about the results? Encourage your family to use the car less, and record its use for another week. Compare the results to the first week.

Technology

Using the Internet, log onto the Web site for your local public transit company. Decide on two destinations to which you would like to travel by bus. Use the appropriate maps and schedule information to plan your trip, including bus route number, departure and arrival times, and fare. Record your information using a suitable format that is easy to read.

Chapter Review

1. In a probability game using the spinner on the right, the spinner was spun 17 times. The results of those spins were

   red: 4   blue: 3   green: 1
   yellow: 2   purple: 7

   a) Express the results in fractions and ratios.
   b) Write three probability statements for the results.

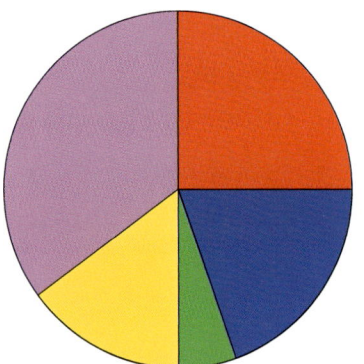

2. A leaky tap will waste enough water in a month for 16 baths. If it keeps dripping, how many baths of water will it waste in 6 months? 10 months? 12 months?

3. The leak gets worse and the number of baths wasted increases by one bath each month. If 16 baths of water are wasted in the first month, 17 in the next, and 18 in the month after that, how many full baths will have been wasted by the twelfth month?

4. If a bath holds 95 L of water, how much water is wasted each month in the first 3 months using the data in question 2? How about the data in question 3?

5. The average Canadian household uses about 500 000 L of water in a year. If 1000 L = 1 m^3, how many cubic metres of water are used?

6. Turning off the tap when you brush your teeth can save at least 1460 L of water in a year. How many litres would a family of three save? How many cubic metres of water is this?

7. In one year, 100 children turning the tap off when they brush their teeth could save 146 000 L of water. How many cubic metres of water is this? Find the amount of litres of water saved and convert it to cubic metres for

   a) 200 children
   b) 300 children
   c) 400 children

Chapter 9: Energy by the Numbers

8. Look at the following circle graphs:

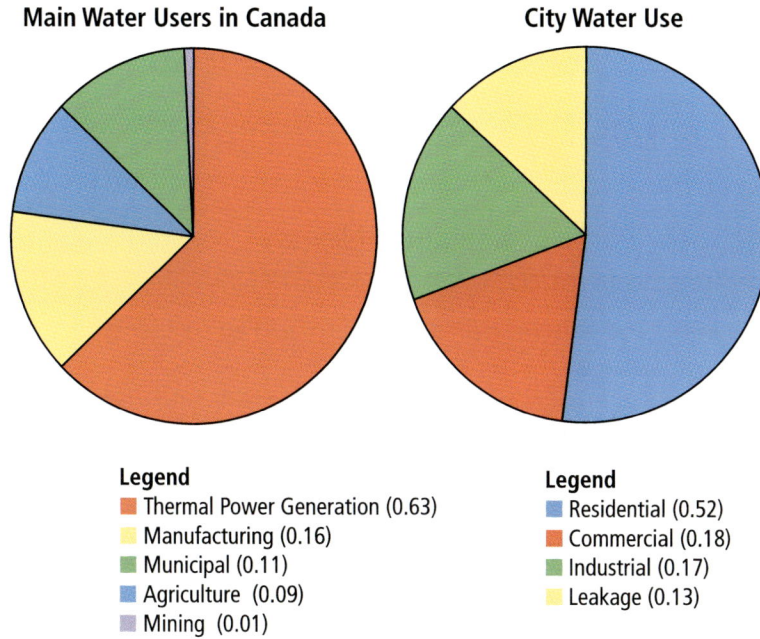

Main Water Users in Canada

Legend
- Thermal Power Generation (0.63)
- Manufacturing (0.16)
- Municipal (0.11)
- Agriculture (0.09)
- Mining (0.01)

City Water Use

Legend
- Residential (0.52)
- Commercial (0.18)
- Industrial (0.17)
- Leakage (0.13)

a) Write each decimal as a fraction.

b) Write a probability statement for $\frac{16}{100}$.

c) Write a probability statement for $\frac{52}{100}$.

d) Write a probability statement of your own.

9. Answer the following questions using the data from the table below.

Typical Water Prices Around the World, 1998 $/1000 L			
Canada	$1.02	Ireland	$1.78
United Kingdom	$3.07	Belgium	$3.75
United States	$1.40	Sweden	$2.61
Netherlands	$3.26	Germany	$4.47
Norway	$1.21	Finland	$2.77
France	$3.37	Australia	$4.09

Chapter Review

a) What is the difference in price between Canada and Australia?

b) If one Canadian uses 343 L of water per person in a day, how much does a day's supply cost?

c) If one Australian uses 112 L of water per person in a day, how much does a day's supply cost?

d) Find the difference in cost between Norway and Belgium. Which country pays more? By how much?

e) If a person in France uses 150 L of water per person in a day, how much does a day's supply cost?

f) If a person from Sweden uses 200 L of water per person a day, how much does a day's supply cost?

g) Compare the cost of water for a Canadian with that of each of the countries in the table. Which country pays the most? By how much?

10. Graph the data in the table from question 9. You may use a line graph or a bar graph. Explain the reason for your choice.

DATA MANAGEMENT AND PROBABILITY
NUMBER SENSE AND NUMERATION
PATTERNING AND ALGEBRA

Chapter Wrap-Up

You are now at the end of Chapter 9. You have analyzed all types of data about energy use. You have looked at tables of energy facts and figures.

As a class, brainstorm a list of what you have learned about the conservation of energy.

How do you feel about energy consumption and conservation? How has the information that you have learned affected your thinking?

Prepare and Present Your Report

Use the ideas mentioned by your classmates and in your discussion. Work with a partner. Review the data about energy consumption presented in this chapter and unit. Prepare a report that outlines ways to lower energy consumption and costs. Use data to support your ideas. Use graphs, charts, and probability statements to present your information. Think of a title for your presentation to help your classmates remember what you said. Then present your report to your class in one of these ways:

- a TV news reports
- a newspaper article
- a TV commercial
- a magazine or billboard advertisement

Understand Other Presentations

Think of three questions to ask classmates based on their presentations.

Technology
Present your report using a multimedia application.

Chapter Wrap-Up **427**

Problems to Solve

Here are some more fun problems for you to solve. For all of these problems, you get to choose a problem-solving strategy.

Problem 15

The Human Tower

STRATEGY: YOUR CHOICE

OBJECTIVE: Estimate long lengths and check your estimates

Problem

Approximately how many Grade 5 students would have to stand on one another's shoulders to be as tall as the CN Tower? First estimate the answer, then calculate it. Use a strategy of your choice to solve this problem.

Tip

Problem-Solving Steps
1. Understand the problem.
2. Pick a strategy.
3. Solve the problem.
4. Share and reflect.

Tip

Pretend all Grade 5 students are the same height.

Height of CN Tower: 553.3 m

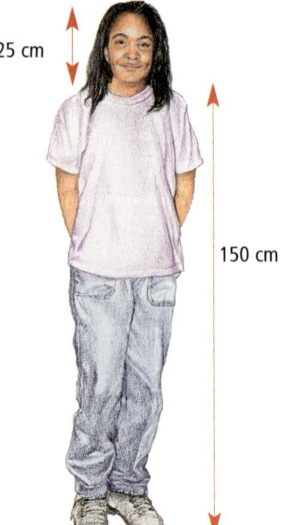

25 cm

150 cm

Chapter 9: Energy by the Numbers

Reflection

1. What did you know about the problem before you began to solve it?

2. What did you need to figure out?

3. Why was it necessary to show in the picture that the head and neck of the Grade 5 student was 25 cm?

4. Explain how fractions or decimals are part of this problem. How can you use fractions and decimals to explain how much shorter the students' height was than the height of the CN Tower?

5. Do you think your strategy was a good choice? Why or why not?

Extension

1. How many Grade 5 students would have to stand on one another's heads to be as tall as the Calgary Tower?

2. How is this Extension question similar to and different from the original problem?

Height of Calgary Tower: 190.9 m

Problems to Solve 429

MEASUREMENT

Problem 16

Capacity and Volume

STRATEGY: YOUR CHOICE

OBJECTIVE:
Relate the volume of an irregular three-dimensional figure to its capacity

Problem

A group of students was determining the capacity and volume of irregular shapes by measuring the amount of water the shapes displaced.

Shape	Amount of Water Displaced (mL)	Capacity (mL)	Volume (cm³)
1	15	15	■
2	13	13	13
3	■	■	■
4	■	■	■

The total volume of all the shapes was 71 cm³. For Shape 3, what could have been the amount of water displaced, the capacity, and the volume? For Shape 4, what could have been the amount of water displaced, the capacity, and the volume? What is the volume for Shape 1? Use a strategy of your choice to solve this problem. Explain any pattern you identify.

430 Chapter 9: Energy by the Numbers

Reflection

1. What information did you have about the problem before you began to solve it?

2. What did you need to figure out?

3. How are capacity and volume linked?

4. Which strategy did you use? Do you think this strategy was a good choice? Why or why not?

5. Compare your results with another classmate's. Are the results different? Why?

Extension

1. If 4 shapes had a total volume of 56 cm^3, and they each had different volumes, what might each of their volumes be? What would each of their capacities be?

2. Share your results with a classmate.

Problems to Solve

Problem 17
Go With the Flow

STRATEGY: YOUR CHOICE

OBJECTIVE:
Solve problems involving whole numbers

Problem

1. The numbers in each box of the flow chart are part of an equation. You need to figure out what the number is in the Start box to make the equation work. Use a strategy of your choice to solve this problem.

2. Create as many of your own equations as you can that equal 120, using addition, subtraction, multiplication, and division. Share your equations with another classmate. Cover one number in each equation and ask your partner to figure out what it is.

Start: ?
↓
÷ 4
↓
+ 29
↓
× 2
↓
= 120

Reflection

1. What did you need to figure out?
2. What did you do first, second, third, and fourth to figure out the flow-chart combination?
3. What strategy did you use to solve each of the problems? Why?
4. How can you check your answers?

Extension

Figure out a slightly different equation for the flow chart. The starting number should stay the same and the answer should still be 120, but change at least two of the steps in the flow-chart boxes.

Journal

Which of the problem-solving strategies that you have learned is your favourite? Explain why.

Chapter 9: Energy by the Numbers

PATTERNING AND ALGEBRA

Problem 18

How Old Is Grandma?

STRATEGY: YOUR CHOICE

OBJECTIVE:
Count by eights and twelves

Problem

Jamar wants to find out the age of his grandmother. To give him a hint, Jamar's grandmother tells him that her age is a multiple of both 8 and 12. What could his grandmother's age be? Figure out all the possibilities. Use a strategy of your choice to solve this problem.

Reflection

1. Rewrite the problem in your own words.
2. What strategy did you use? Why?
3. Was this strategy a good one to use? Why or why not?
4. Was there more than one answer to this problem? If so, which answers make sense? Explain.
5. Share your results with another classmate.

Extension

Work with a partner. Ask your partner to think of a number between 20 and 100 without telling you what it is. Write a series of questions that you could ask that would help you figure out the number. All the questions must have yes or no answers. Now ask your partner the questions, and try to guess the number. Then let your partner have a turn figuring out a number that you think of.

Problems to Solve 433

Celebrating Math

Congratulations! You have learned a lot of new math this year, and now it's time to use what you have learned to celebrate. You will prepare for a class bake sale. There are six fun activities that let you practise the math skills you have learned. Before you begin Celebrating Math, brainstorm a list of math ideas that you learned this year. Look back through this book and your notebook for ideas. Think about the most challenging and easiest math concepts that you learned. Share your thoughts with a classmate.

At the beginning of the year, you were asked, "If 100 is the answer, what is the question?" You have learned a lot about math since then! Think about that question again. How can you answer it differently now? Think of similar problems and try them with your classmates, using all of the math knowledge you have learned this year.

NUMBER SENSE AND NUMERATION

Lesson 1
Working With Division

BAKE SALE LOG

PLAN:
You will use your division skills to separate cookies into bags.

DESCRIPTION:
Selling baked treats like cookies, cupcakes, and pastries is a great way to raise money for various school events. Today you will begin preparing for your school's bake sale.

Get Started

Make a list of what you know about division. Give examples to show your thinking.

Build Your Understanding

Cookie Division

You Will Need
- base-ten blocks

Five students in your class brought cookie sheets filled with cookies to school. Your task is to help the students divide the cookies into gift bags.

Lesson 1: Working With Division

Use the base-ten blocks to help you figure out these problems.

1. How many cookies are there altogether?
2. How many bags of cookies will there be if there are 13 cookies in each bag? How many cookies would be left over?
3. How many bags of cookies will there be if there are 16 cookies in each bag? How many cookies would be left over?
4. How many bags of cookies will there be if there are 21 cookies in each bag? How many cookies would be left over?

What Did You Learn?

1. Explain how you figured out how many cookies would be in each bag.
2. How are your answers for questions 2 to 4 in Build Your Understanding related?

Practice

1. Use division to divide the numbers shown in the cookies below by 8.

A 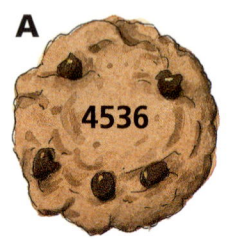 4536
B 7062
C 1088
D 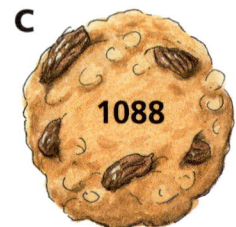 3100
E 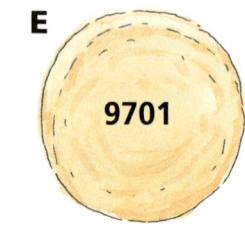 9701

2. Use a calculator to check your answers.

436 Celebrating Math

MEASUREMENT

Lesson 2

Perimeter, Area, and Volume

BAKE SALE LOG

PLAN:
You will find the perimeter, area, and volume of several cakes.

DESCRIPTION:
There are many different shapes and sizes of cakes at the bake sale.

Get Started

Draw a picture of a rectangle, and then measure its sides. Now explain the rule for finding its perimeter and area.

Build Your Understanding

Find Perimeter, Area, and Volume

Several of your classmates brought cakes for the bake sale. The cakes were made in the shapes shown to the right. Your job is to find the perimeter, area, and volume of each one. Make a chart to record your answers.

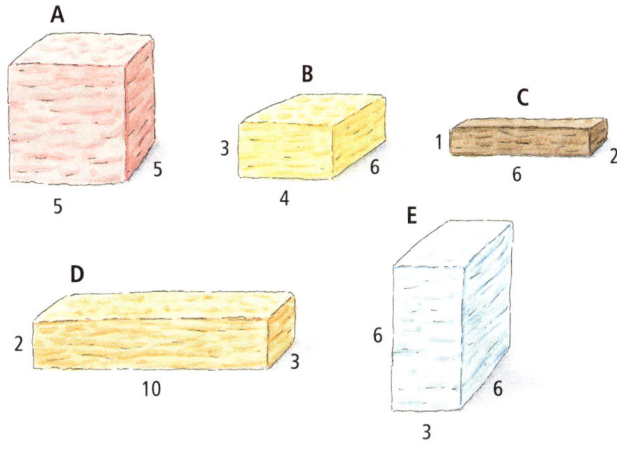

Tip
Creating a net for each three-dimensional shape will help you find the perimeter and area.

Lesson 2: Perimeter, Area, and Volume 437

What Did You Learn?

1. Which cake had the longest perimeter?
2. Which cake had the least volume?
3. Which cake had the greatest area?
4. Which cake would serve the most people? Explain why.
5. Arrange the cakes in order from the one with the largest area to the one with the smallest area.
6. Is it possible for two cakes to have the same perimeter but different areas? Explain.

Practice

You Will Need
- linking cubes

Design a cake using linking cubes, and then find its perimeter, area, and volume.

Extension

Find the area of these cake shapes. Explain how you got your answers.

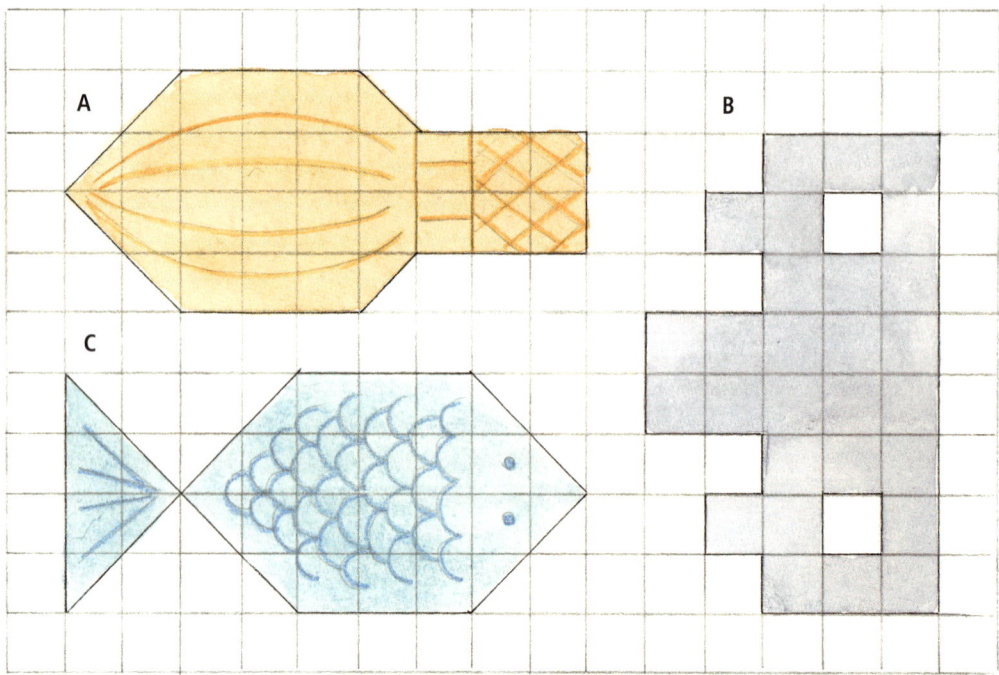

NUMBER SENSE AND NUMERATION

Lesson 3
Fractions and Decimals

BAKE SALE LOG

PLAN:
You will name pies using fractions and decimals.

DESCRIPTION:
Cherry, blueberry, peach, and apple are all flavours of pies that might be found at a bake sale. Which one is your favourite?

Get Started

In your journal, draw pictures to show what you know about fractions and decimals.

Build Your Understanding

Fractions

Some of your classmates brought pies for the bake sale that were filled with two different fruits. Write two fractions and two decimals to describe the contents of each pie.

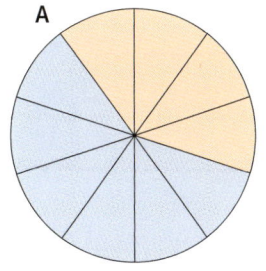

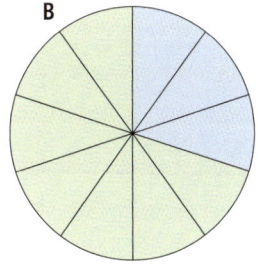

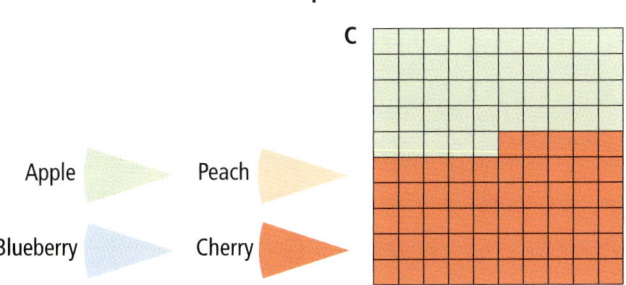

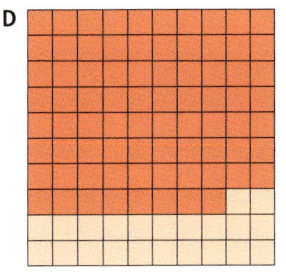

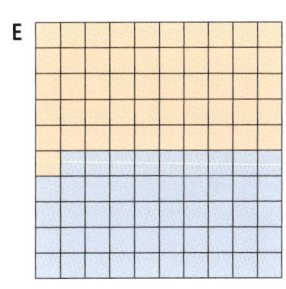

Lesson 3: Fractions and Decimals   439

What Did You Learn?

1. What is the difference between calculating the fraction of cherries in one pie and calculating the fraction of pies with cherries? Which fraction would be part of a whole, and which fraction would be part of a set? Explain how you know.

2. Write fractions that describe the total number of pieces for each pie flavour. For example, $\frac{58}{320}$ are blueberry.

3. Which flavour has the most pieces?

4. Which flavour has the fewest pieces?

Practice

You Will Need
- base-ten blocks

1. Use base-ten blocks to make fractions of a thousand block. Challenge a classmate to name the fraction and the decimal. Take turns.

Extension

What is the place value of the underlined digits?

2. <u>7</u>6 432
3. 9<u>0</u> 081
4. 9<u>6</u> 800
5. 38 9<u>8</u>6
6. 14 08<u>2</u>
7. 67 <u>9</u>75

8. Round each number from questions 2 to 7 to the nearest ten thousand.

440 Celebrating Math

Lesson 4
Exploring Large Numbers

BAKE SALE LOG

PLAN:
You will work with large numbers.

DESCRIPTION:
Sometimes there are special draws or contests at bake sales.

Get Started

1. Copy the chart below into your notebook and complete it, recording what happened to get from the start number to the finish number. The first row has been done for you.

Start Number	Finish Number	What Happened?
56 789	98 790	42 001 was added to the start number.
36 984	58 888	
69 953	96 651	
46 780	46 870	

Create another chart like this one using your own start and finish numbers. Exchange your chart with a classmate.

Build Your Understanding

Estimate and Check

You Will Need
- clear glass or plastic container, any size
- small cubes

Work with a partner.
1. Fill your container with cubes.
2. Look carefully at the container and estimate the number of cubes in it.
3. Count to check your estimate.
4. Repeat steps 1 to 3.

What Did You Learn?

1. How did you arrive at your estimate the first time? How did you arrive at your estimate the second time?

2. How far away was your estimate from the actual amount the first time? How far away was your estimate from the actual amount the second time?

3. How many cubes would there be if you had 25 times more than the actual amount you had when you filled the container the first time? the second time?

Practice

Write these numbers in words and then in expanded form.

1. 67 894
2. 10 560
3. 83 073
4. 99 999

A Math Game to Play

You Will Need
- egg carton
- felt pen
- six bread tags

How to Play

Work with a partner or in a small group.

1. Record a number of your choice in each of the 12 parts of the egg carton.

2. Label the bread tags 1, 10, 100, 1000, 10 000, 100 000.

3. Put the bread tags in the egg carton and shake it.

4. Open the egg carton and multiply each bread tag number by the number written on the egg carton section that it landed in.

5. Add up all of your products to find one large sum.

6. Repeat this game using different numbers in your egg carton sections.

Celebrating Math

DATA MANAGEMENT AND PROBABILITY

Lesson 5
Counting Money

BAKE SALE LOG

PLAN:
You will calculate how much money was made at the bake sale.

DESCRIPTION:
The money made at a bake sale is usually for a good cause.

Get Started

The graph shows how much money four people spent at the bake sale.

1. Write as many facts about the graph as you can think of.

2. How much money did all four people spend in total?

3. What is the mean amount of money shown on the graph? Show your work.

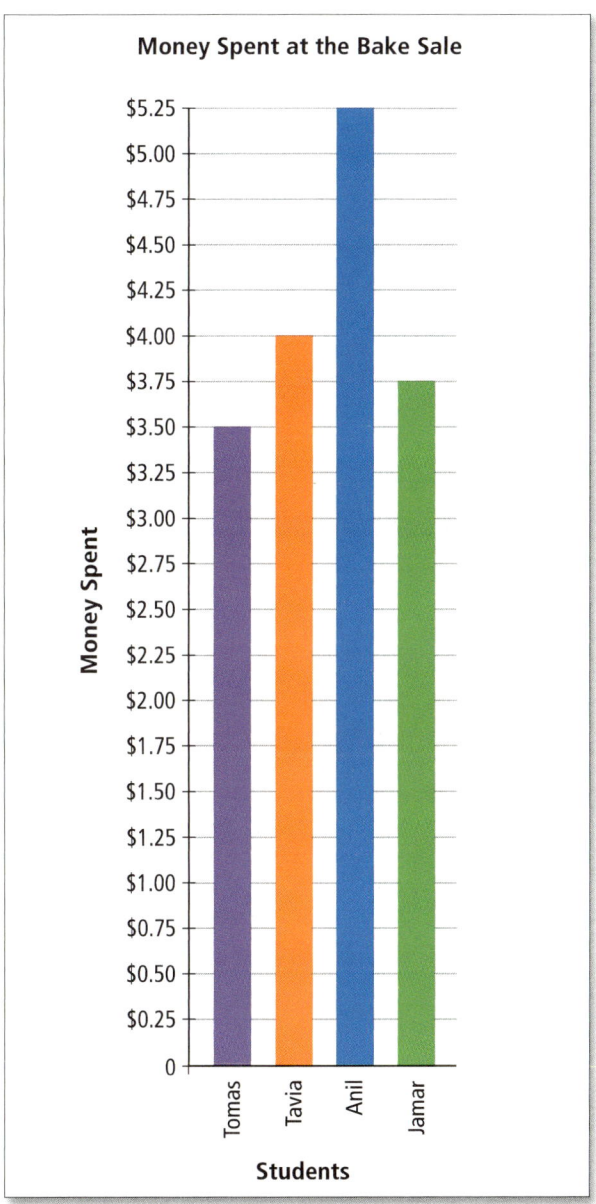

Lesson 5: Counting Money

Build Your Understanding

Analyze Proceeds

You Will Need
- play money

The chart shows what each student bought at the bake sale.
Copy and complete the chart in your notebook.

Name	Bought	Amount	Cost	Total Cost	Paid	Change
Jessica	donuts	2 dozen	$1.12 per dozen		$5.00	
Noma	cakes	4 cakes	$4.37 each		$20.00	
Shefali	pies donuts	7 pies 1 dozen	$3.89 each		$40.00	
Kenny	cookies	6 bags of 3 cookies	$0.34 per cookie		$10.00	
Jamar	banana bread cakes	5 loaves 2 cakes	$2.90 per loaf		$30.00	

What Did You Learn?

1. According to the chart, how much money was raised altogether?
2. Who spent the most?
3. Who spent the least?
4. Arrange the items bought in order from most expensive to least expensive.

Practice

A Math Problem to Solve

1. Look at your chart from this lesson's Build Your Understanding. Record the different combinations of coins that could have been given to each person for their change.

Extension

2. Use the information on the chart to make a graph. Make sure you label your graph correctly, and don't forget to include a title.

Celebrating Math

Planning a Class Bake Sale

You have reached the end of Celebrating Math. You have used your math skills to investigate and solve several math questions about a bake sale. You have also learned that math is a big part of a bake sale, and that without math a bake sale would not be possible.

To celebrate the math you learned in this unit and throughout this year, write a plan for a bake sale item that you will sell. Here are a few questions to consider:

1. What would you like to make for the bake sale?

Technology
With your teacher's permission, research the recipe for your bake sale item using the Internet.

2. What ingredients are needed?

3. How much of each ingredient will you need?

4. How will you decorate and package your item?

5. How much will it cost to make your item?

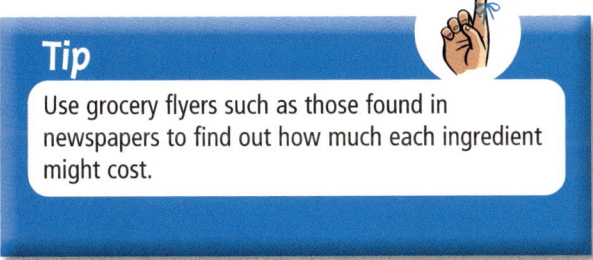

Tip
Use grocery flyers such as those found in newspapers to find out how much each ingredient might cost.

6. How much would you sell your item for? How much profit could you make on this item?

7. What could the money you raise be used for?

8. How will you advertise or display your bake sale item?

When you have answered these questions, share your plan with the teacher. Then, write an article about the bake sale for your school's newsletter. Now you are ready for the bake sale! Ask an adult in your family to help you make your bake sale item at home. When it is ready for sale, bring it to school. Your teacher will help you set up your class bake sale. Have fun!

Celebrating Math

Glossary

acute angle An angle that measures less than 90 degrees

Example: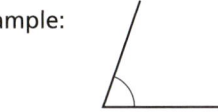

addends The numbers you add together in addition calculations

algebra A branch of mathematics that uses expressions, variables, constants, and other symbols to make mathematical statements

algebraic expression One or more variables and sometimes numbers and operation symbols; for example, y, $3y$, and $2x + 7 = y$ are algebraic expressions

algorithm A set of rules or procedures for doing computations such as addition, subtraction, multiplication, or division. For example, the addition algorithm is a set of rules for finding the sum of two or more numbers.

A.M. Before noon (from 12:00 A.M. [midnight] to 11:59 A.M.)

analog clock A clock that indicates time through the position of its hands

anemometer An instrument used to measure wind speed

angle Line segments that share an endpoint. We measure angles in degrees (°).

Example:

area The amount of surface inside a two-dimensional shape. Area is measured in square units: mm^2, cm^2, dm^2, m^2, km^2. Square units are units that measure equally on all four sides.

Example: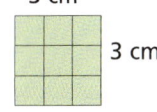

area = 9 cm^2

axis One of the intersecting number lines on a graph. Two or more of these lines are called axes.

bar graph A diagram that uses horizontal or vertical bars to show data

Example: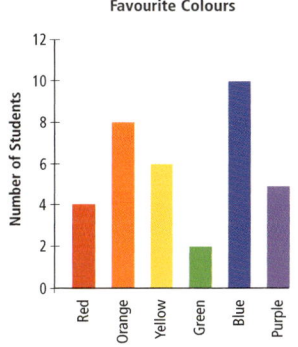

barometer An instrument used to measure air pressure

base A surface on which a figure can stand

Examples: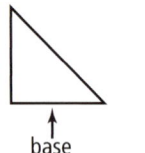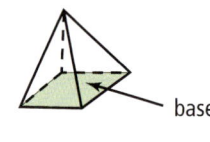

broken-line graph A graph that displays data by showing points joined by lines

Example: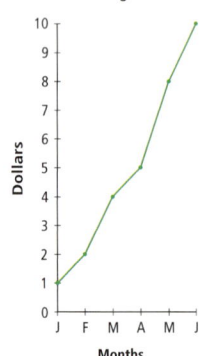

capacity The amount a container can hold when it is filled

centi A prefix meaning one hundredth ($\frac{1}{100}$)

centimetre One hundredth of one metre or 0.01 m. The metric symbol for centimetre is cm.

circumference The distance around, or perimeter of, a circle

climate The usual weather pattern a place has over a long period of time

clustering A strategy that can be used to estimate the sum of numbers that cluster around the value of one number. For example, the numbers 44, 45, 54, and 55 cluster around 50. You can estimate that 44 + 45 + 54 + 55 ≈ 200.

composite number A number that has factors besides 1 and itself. For example, the number 6 has four factors: 1, 2, 3, and 6.

congruent Exactly the same size and shape

constant A term in an algebraic expression that does not change. For example, in $x + 3$, the constant is 3.

cube A three-dimensional solid with six congruent square faces

Example: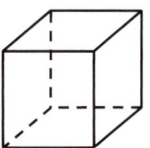

D

decimal number A number written in the decimal system. For example, 0.5, 0.6, or 1.3.

decimetre One tenth of a metre or 0.1 m

denominator The bottom number in a fraction

Example:

$\frac{3}{4}$ ← denominator

diagonal A line segment that joins two opposite verticies on a two-dimensional shape

Example: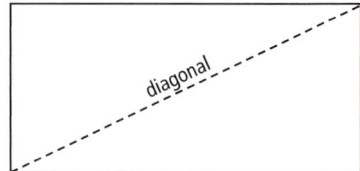

Math Everywhere 5

diameter A line segment that joins two points on the circumference of a circle and passes through the centre of the circle

Example:

A •———•———• B   diameter: $\overline{AB}$

digit A symbol used to record a number; the symbols 0, 1, 2, 3, 4, 5, 6, 7, 8, and 9 are digits

dimension A measurement such as the height, width, length, or depth of an object

dividend The number being divided in a division calculation. For example, in 36 ÷ 6, the dividend is 36.

divisor The number you are dividing the dividend by. For example, in 15 ÷ 3, the divisor is 3.

duration How long something takes or lasts

E

edge The line segment where two faces of a figure meet

Example: — edge

equal angles Angles that have the same measure

equal to (=) This symbol means "is equal to"
Example: 5 x 5 = 25

equation A mathematical statement that has equivalent terms on either side of the equal sign (=); for example, 2 x 3 = 6

equilateral triangle A triangle in which all sides are the same length

Example: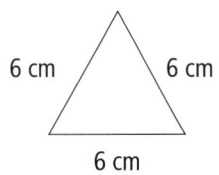
6 cm, 6 cm, 6 cm

equivalent Equal to; the same amount

equivalent fractions Fractions that represent the same part of a whole or a group. $\frac{1}{2}$, $\frac{2}{4}$, $\frac{3}{6}$, $\frac{4}{8}$, $\frac{5}{10}$, and $\frac{6}{12}$ are all equivalent fractions.

estimation strategies Mental mathematical strategies used to obtain approximate answers

even number A whole number with 2 as a factor. The number can be divided by 2.

expanded form A way of writing numbers that shows the value of each digit
Example: 6352 = 6000 + 300 + 50 + 2

expanding pattern A pattern in which the elements of the pattern increase with each extension

F

face The flat side of a figure

Example: — face

Glossary **449**

factor A number that divides evenly into another number or expression. For example, 3 is a factor of 6.

formula A set of symbols, such as letters and numbers, used to give a general rule. For example, the formula for finding the perimeter of a rectangle is $P = 2l + 2w$; the formula for finding the area of a rectangle is $A = l \times w$.

frequency The number of times an event or item occurs

frequency table A table showing the number of times an event occurs

geoboard A piece of wood or plastic that has rows of pegs or nails evenly spaced

graph A visual way to represent, or display, data

greater than (>) A symbol meaning "is greater than"
Example: 30 > 25

grid paper Paper marked off in equal-sized squares. For example, 1 cm x 1 cm, or 1 cm².

hectare A unit of measurement equal to one square hectometre. A hectometre equals 10 000 m.

height The distance from the base of something to its top; how tall something is
Example:
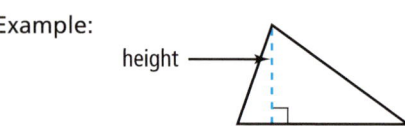

heptagon A polygon with seven sides; *hepta* means seven

hexagon A polygon with six sides; *hexa* means six
Example:
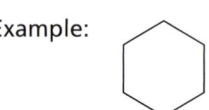

hygrometer A device used to measure the air's humidity

improbable Not likely to happen

interval A space between two points, such as between the start and end of a period of time, a pair of given numbers, or two points on an axis in a graph

irregular shape A shape with unequal side and angle measures

isosceles triangle A triangle that has two sides of equal length
Example:
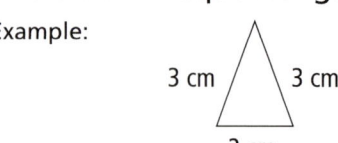

kilo A prefix meaning a group of 1000

kilometre A unit for measuring distance equal to 1000 m. The metric symbol for kilometres is km.

kilowatt hour A unit for measuring electricity equal to 1000 W of electricity used in one hour. The metric symbol for kilowatt hour is kWh.

legend A feature on a pictograph that explains what each symbol represents, or stands for

length The distance from one end of something to the other end; how long something is

less than (<) A symbol meaning "is less than"
Example: 25 < 30

line graph A graph that uses a line or lines to show data or information and its relationships
Example:
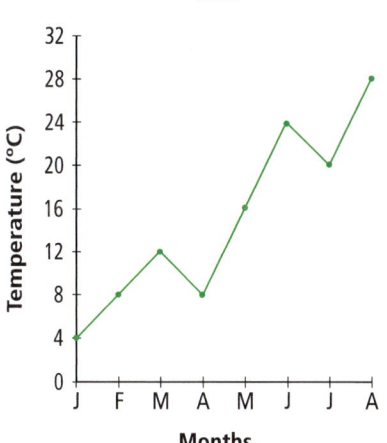

line segment A part of a line between two points on the line
Example: A———B

mean The average of a set of numbers. To calculate the mean, add up all the numbers and divide by the number of choices given. For example, if you had 1 item using 3 energy forms, 1 item using 2 energy forms, and 1 item using 1 energy form, the mean would be 2 energy forms (3 + 2 + 1 = 6, 6 ÷ 3 = 2).

megawatt A unit for measuring electricity equal to 1000 kW or one million watts. The metric symbol for megawatt is MW.

metre A unit of measurement used to measure length. The metric symbol for metre is m.

milli A prefix meaning one thousandth ($\frac{1}{1000}$)

millimetre One thousandth of one metre or 0.001 m. The metric symbol for millimetre is mm.

mixed number A whole number and a fraction together. For example, $2\frac{1}{3}$ is a mixed number.

mode The most frequent number in a group. If you have 5 items using heat energy, 5 items using sound energy, 3 items using motion energy, and 1 item using light energy, the mode is 5.

multiple The product of a whole number with another whole number. For example, 12 is a multiple of 6 because 2 x 6 = 12.

multi-step problem A problem in which you must use at least two calculations to find the solution

Glossary

net A pattern that can be folded to make a three-dimensional figure

Example:

nonagon A polygon with nine sides; *nona* means nine

number line A line divided into equal parts with dividing lines labelled with numbers

Example:

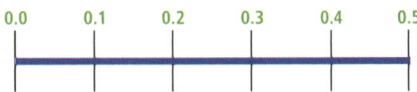

numerator The top number in a fraction

Example:

$\frac{3}{4}$ ← numerator

obtuse angle An angle that measures more than 90 degrees but less than 180 degrees

Example:

octagon A polygon with eight sides; *octa* means eight

Example:

odd number A number that does not have 2 as a factor. The number cannot be divided by 2.

operation Addition, subtraction, multiplication, or division

parallelogram A quadrilateral whose opposite sides are parallel

Example:

pattern Something, such as designs or numbers, repeated in regular ways

pentagon A polygon with five sides; *penta* means five

Example:

perimeter The distance around a figure

perpendicular Meeting at a right angle, as lines or line segments do in the corner of a square

Example:

pictograph A graph that shows data using pictures as symbols

Example:

How We Get to School	
Walk	★ ★ ★
Ride a Bike	★ ★ ★ ★
Ride a Bus	★ ★ ★ ★ ★ ★
Ride in a Car	★ ★

Legend: Each ★ = 10 students

Math Everywhere 5

place value The value given to the place in which a digit appears in a number. For example, in 12 683, 1 is in the ten-thousands place, 2 is in the thousands place, 6 is in the hundreds place, 8 is in the tens place, and 3 is in the ones place.

plane A flat or level surface that extends without end in all directions
Example:
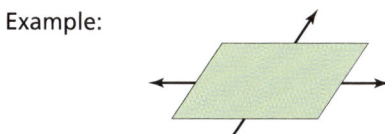

P.M. After noon (from 12:00 P.M. [noon] to 11:59 P.M.)

polygon A closed shape that is formed by three or more line segments. Some examples of polygons are triangles, quadrilaterals, and pentagons.

population The total number of people in a specific area

precipitation Any form of water that falls from the sky, including rain, snow, sleet, and hail

prediction A reasonable guess as to the outcome of an event

prefix A group of letters added to the front of a word that changes its meaning
centi: A prefix meaning $\frac{1}{100}$, as in centimetre
deci: A prefix meaning $\frac{1}{10}$, as in decilitre
kilo: A prefix meaning 1000, as in kilometre
milli: A prefix meaning $\frac{1}{1000}$, as in millilitre

prime number Any number other than 1 whose only factors are 1 and itself. For example, 5, 7, 11, 13, 17, and 19 are prime numbers.

prism A three-dimensional figure with two faces that are congruent and parallel and other faces that are parallelograms
Example:
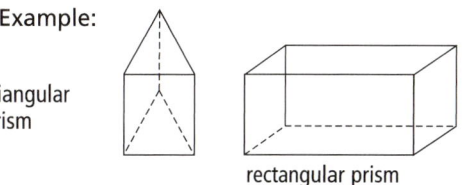
triangular prism
rectangular prism

probability A number statement that tells how likely it is that an event will take place

probability statement A way of saying your predictions. If you think there's a 1 in 5 chance of an event happening, you say there's a 1 in 5, $\frac{1}{5}$, or 1:5 probability.

probable Likely to happen

product The answer in a multiplication calculation

pyramid A solid figure with a polygon base and triangular faces that meet at a common point
Example:
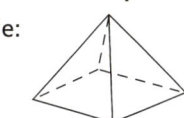

 Q

quotient The answer in a division calculation. For example, in 8 ÷ 4 = 2, the quotient is 2.

Glossary

R

rain gauge An instrument used for measuring rainfall

rate A relationship between quantities measured in different units. For example, 35 km/h means that something will travel 35 km in one hour.

reflection (flip) A transformation (movement) of a figure by flipping it over a mirror line

Example: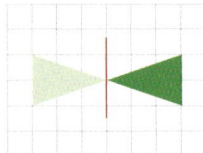

repeating pattern A pattern in which the same elements repeat

right angle A square corner angle. A right angle is 90 degrees.

Example:

rotation (turn) A transformation (movement) of a figure around a fixed point

Example: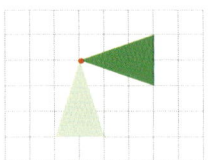

rounding A strategy that can be used to give an approximate number for an exact number to a nearest place-value position. For example, 5769 to the nearest hundred is 5800.

S

sample Part of the whole group. When determining probability, the larger the sample, the more accurate the results.

scalene triangle A triangle with three sides of different lengths

Example: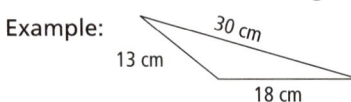

sequence A succession of things, including numbers, that are connected in some way; for example, the sequence of numbers 1, 2, 2, 3, 3, 3, 4, 4, 4, 4, ...

solid figure A three-dimensional shape

standard form A way of writing numbers in which each digit has a place value according to its position in relation to other digits; for example, 8.73

strategy A plan to help you solve problems

sum The answer in an addition calculation

survey A tool for collecting information by asking people questions or interviewing them

symbol Something, such as a letter, picture, or number, that represents, or stands for, something else

symmetry Two congruent sides divided by a line

table A useful tool for organizing and presenting information

tally chart A chart that uses tally marks to count data and record how frequently something happened

tangram An ancient Chinese puzzle made from a square cut into seven pieces: two large triangles, one medium-sized triangle, two small triangles, one square, and one parallelogram

terms The parts of an algebraic expression separated by addition or subtraction signs. For example, 4x + 5 has two terms, 4x and 5.

tessellation A repeating pattern of closed figures that covers a surface with no gaps and overlaps
Example:
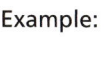
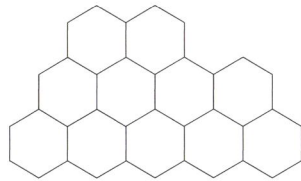

thermometer An instrument used for measuring temperature

three-dimensional Having three dimensions—length, width, and height
Example:
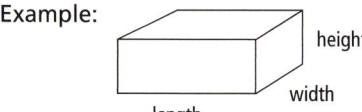

tile To cover a plane surface with a combination of different shapes with no gaps or overlaps

transformation A movement that does not change the size or shape of a figure

translation (slide) A transformation (movement) of a figure along a straight line
Example:
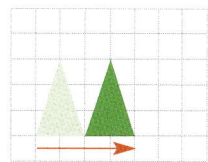

tree diagram A branching graph showing the possible outcomes in a probability experiment

two-dimensional Having two dimensions—length and width, but not height
Example:
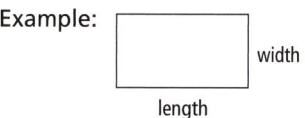

variable A letter or symbol used to represent a number. A variable changes; for example, in the formula 2*l* + 2*w*, *l* and *w* are variables.

vertex The point where two line segments or arms of an angle meet. If there is more than one vertex, they are called vertices.
Example:
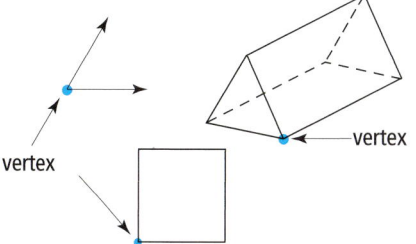

volume The amount of space an object takes up. A measurement unit of volume is cubic centimetres, or cm^3.

W

watt A unit used to measure the rate of electricity use. The metric symbol for watt is W. The lower the watts, the less electricity used. The amount of energy used is measured in watts per hour (watt hours or Wh) or kilowatt hours (kilowatts per hour or kWh). 1 kW = 1000 W

wattage The amount of electricity

weather The condition of the atmosphere at a certain time and place. For example, it might be hot, cold, rainy, or snowy.

weather vane An instrument used for measuring wind direction

whole number The numbers 0, 1, 2, 3, 4, 5, 6, and so on

windsock An instrument used for measuring wind direction

written form A way of writing numbers in words. For example, 6352 in written form is "six thousand three hundred fifty-two."

Problem-Solving Strategies

Here are some helpful strategies that you can use to solve the problems that appear in *Math Everywhere*.

Problem solving is a very important part of math. Problem solving allows you to practise the math skills you have learned.

Sometimes one strategy might not help you find the solution. If that happens, try another strategy. Trying different ways to find an answer is a part of learning.

1. Act out the problem to help visualize the solution.
2. Draw a picture to help you solve the problem.
3. Use objects like counters or play money to find the solution.
4. For large numbers, you can make a guess and then check to see if your guess is correct.
5. Begin with the information from the end of the problem and work backwards to find the solution.
6. For complicated problems, you can reduce large numbers to smaller numbers or reduce the number of items in the problem.
7. You can use concrete materials to build a model.
8. Looking for a pattern can help you solve many different kinds of problems.
9. You can make a table or a chart to help you organize information and find patterns in the data.
10. Making a list helps you organize information, too.

Here are some important steps to follow for all problems, no matter which strategy you use.

1. **Understand the problem:** Rewrite the problem in your own words. If you can, draw a picture of the problem. List or highlight important numbers or words.

2. **Pick a strategy:** You will be learning about many different problem-solving strategies throughout the year; for example, "Act It Out," "Draw a Picture," "Use Objects," and "Guess and Check."

3. **Solve the problem:** Use a strategy to solve the problem. Describe all steps using math words and/or symbols. Try a different strategy if you need to. Organize the results using a diagram, model, chart, table, or graph.

4. **Share and reflect:** Did the strategy you picked work? Would a different strategy also work? Does your solution make sense? Could there be more than one answer to the problem? How did other people in your class solve the problem?

Acknowledgements
Photographs

t = top, b = bottom, l = left, r = right, m = middle

Pg 20: © Catherine Jones/SuperStock; 21: (l) Ivy Images, (r) CP/St. John's Telegram(Keith Gosse); 22: Bill Ivy/Ivy Images; 26: CP(Jacques Boissinot); 29: Ivy Images; 34: Courtesy of CN Tower; 37: (t) Lone Pine Photo, (b) O. Bierwagen/Ivy Images; 42: (t) CP(Tim Clark), (b) A. Pichette/CIS/Ivy Images; 46: CP/St. John's Telegram(Keith Gosse); 50: Courtesy of Atmospheric Monitoring Division, Meteorological Service of Canada – Pacific and Yukon Region; 53: CP(P. Anil Kumar); 56: CP/Winnipeg Free Press(Ken Gigliotti); 60: Lone Pine Photo; 63: © Tom Bean/CORBIS/MAGMA; 68: (l) © AFP/CORBIS/MAGMA, (r) CP/ Saskatoon Star Phoenix(Greg Pender); 69: (t) CP(Andrew Vaughn), (b) CP(Fred Chartrand); 73: *Satellite images*, http://www.weatheroffice.ec.gc.ca/satellite/index_e.html, State of the Environment Reporting, Environment Canada 2003. Reproduced with the permission of the Minister of Public Works and Government Services, 2003; 76: *Satellite images*, http://www.weatheroffice.ec.gc.ca/satellite/index_e.html, State of the Environment Reporting, Environment Canada 2003. Reproduced with the permission of the Minister of Public Works and Government Services, 2003; 78: *Environment Canada's Web site*, http://www.ec.gc.ca/envhome.html, Environment Canada, 2003. Reproduced with the permission of the Minister of Public Works and Government Services, 2003; 85: Bill Ivy/Ivy Images; 94: (t) CP/Winnipeg Free Press(Phil Hossack), (bl) CP(Chuck Stoody), (br) W. Towriss/Ivy Images; 95: Source: *The Climates of Canada*, Cat. No. En56-1/1990E, Minister of Supply and Services Canada 1990; 97: Adapted from Munich Re (1996), from *Extreme Weather and Climate Change*, Minister of Supply and Services Canada 1998; 98: CP/Regina Leader-Post (Bryan Schlosser); 102: (l) Bill Ivy/Ivy Images, (r) René Johnston/Toronto Star; 105: Source: *The Climates of Canada*, Cat. No. En56-1/1990E, Minister of Supply and Services Canada 1990; 107: (t) CP(Kevin Frayer); 109: Source: The Climates of Canada, Cat. No. En56-1/1990E, Minister of Supply and Services Canada 1990; 114: Mike Grandmaison; 116: (m) G. Wiltsie/CIS/Ivy Images, (tl) J. DeVisser/Ivy Images, (tr) Greg Locke, (bl) Bill Ivy/Ivy Images, (br) Ron Smid/firstlight.ca; 117: (t) S. O'Neill/Ivy Images, (b) W. Fraser/Ivy Images; 121: William Lowry/Ivy Images; 122: Tom Kitchin/firstlight.ca; 123: © Katherine Karnow/CORBIS/MAGMA; 130: (t) © GoodShoot/SuperStock, (b) © George Hunter/SuperStock; 138: (l) © Roger Allyn Lee/SuperStock, (tr) Fotopic/MaXx Images, (mr) Alan Marsh/firstlight.ca, (mr) Greg Locke, (br) Barrett & MacKay; 139: (l) Mike Grandmaison, (r) William Lowry/Ivy Images; 141: B. M. Wolitski/CIS/Ivy Images; 143: (t) Arjen and Jerrine Verkaik/SKYART, (b) CP(Adrian Wyld); 145 CP(Jacques Boissinot); 147: Bill Ivy/Ivy Images; 151: (t) Bill Ivy/Ivy Images, (ml) W. Fraser/Ivy Images, (mr) W. Fraser/Ivy Images, (b) National Center for Atmospheric Research/University Corporation for Atmospheric Research/National Science Foundation; 166: © Gary Braasch/CORBIS/MAGMA; 168: Max & Bea Hunn/Visuals Unlimited; 170: (t) Bill Kamin/Visuals Unlimited, (b) David Matherly/Visuals Unlimited; 173: (l) Francis E. Caldwell/Visuals Unlimited, (ml) Visuals Unlimited, (mr) © Royalty-free/CORBIS/MAGMA, (r) Ashmolean Museum, Oxford, UK/Bridgeman Art Library; 178: (l) The Granger Collection, New York, (r) © Royalty-Free/CORBIS/MAGMA; 183: (r) © The British Museum; 188: The Granger Collection, New York; 192: Bruce Berg/Visuals Unlimited; 196: Max & Bea Hunn/Visuals Unlimited; 200: © Jonathan Blair/CORBIS/MAGMA; 205: © Geoffrey Clements/CORBIS/MAGMA; 208: (l) The Granger Collection, New York, (r) © Archivo Iconografico, S.A./CORBIS/MAGMA; 215: (t) © Margaret Courtney-Clarke/CORBIS/MAGMA, (m) © Royalty-Free/CORBIS/MAGMA, (b) © Christie's Images/CORBIS/MAGMA; 218: Will Troyer/Visuals Unlimited; 220: CP PHOTO; 221: Bud Nielsen/Visuals Unlimited; 222: © Layne Kennedy/CORBIS/MAGMA; 223: (l) The Granger Collection, New York, (r) © Christie's Images/CORBIS/MAGMA; 227: Courtesy of Beacon Hill Park, Victoria B.C.; 230: Max & Bea Hunn/Visuals Unlimited; 238: Jeff Greenberg/Visuals Unlimited; 239: Valley of the Kings, Thebes, Egypt/Bridgeman Art Library; 260: Ken Martin/Visuals Unlimited; 263: (t) Ivy Images, (b) CP(Frank Gunn); 265: (l) Julian Worker/Visuals Unlimited, (ml) MaXx Images, (m) © Steve Vidler/SuperStock, (mr) Nada Pecnik/Visuals Unlimited, (r) © Dean Conger/CORBIS/MAGMA; 267: (t) Dale Jackson/Visuals Unlimited, (b) © Michelle Garrett/CORBIS/MAGMA; 275: Fotografia, Inc/CORBIS/MAGMA; 279: Julian Worker/Visuals Unlimited; 282: Bill Kamin/Visuals Unlimited; 286–287: Tangrams used by permission of Learning Resources, Inc.; 288: The Granger Collection, New York; 291: Charles Preitner/Visuals Unlimited; 295: National Museum of India, New Delhi, India/Bridgeman Art Library; 302: © Vanni Archive/CORBIS/MAGMA; 303: Stone/Getty Images; 309: Dick Hemingway; 310: Bill Ivy/Ivy Images; 311: O. Bierwagen/Ivy Images; 312: Bill Ivy/Ivy Images; 316: Dick Hemingway; 323: Reproduced with the permission of Her Majesty the Queen in Right of Canada, 2002; 328: Chart source: Reproduced with the permission of Her Majesty the Queen in Right of Canada, 2002; 333: Dick Hemingway; 355: Bill Ivy/Ivy Images; 358: G. Daigle/CIS/Ivy Images; 364: G. Daigle/CIS/Ivy Images; 376: Adapted from Statistics Canada "Primary Energy by Source, Canada, 1871 to 1996 (Percentage of Energy Consumption)", data is based in part on the Statistics Canada publication "Energy Statistics Handbook", Cat. 57-601, 2002, Table 2.9; 392: W. Lowry/Ivy Images; 393: Bill Ivy/Ivy Images; 396: Bill Ivy/Ivy Images; 405: CP(Chuck Stoody); 411: CP(Jonathan Hayward); 414: Bill Ivy/Ivy Images; 415: © Her Majesty the Queen in Right of Canada. All rights reserved. Source: Freshwater Website (www.ec.gc.ca/water), Environment Canada, 2002. Reproduced with the permission of the Minister of Public Works and Government Services, 2002. 417: Dick Hemingway; 418: Table entitled "International Prices for Motor Gasoline in Cents Per Litre (taxes included)", adapted from the Statistics Canada Web site http://www.statcan.ca/english/ads/57-601-XDE/temp4.htm; 428: Dick Hemingway; 429: Troy and Mary Parlee/MaXx Images

Statistics Canada information is used with the permission of the Ministry of Industry, as Minister responsible for Statistics Canada. Information on the availability of the wide range of data from Statistics Canada can be obtained from Statistics Canada's Regional Offices, it's World Wide Web site at http:www.statcan.ca, and its toll-free access number 1-800-263-1136.

Illustrations

Deborah Crowle: pp. 39, 86, 88, 345, 415 (b); Malcolm Cullen: pp. 217, 235 (t), 238, 240, 242, 250; Greg Douglas: pp. 1, 3, 4 (t), 12, 56, 66, 71, 84, 103, 134, 160, 264, 271(b), 295, 306, 308, 314, 319 (b), 321, 323, 337, 356 (t), 360, 368 (b), 371, 373, 375 (b), 380, 382 (b), 384, 386, 390, 394, 402 (t), 415 (t), 416, 421 (b), 427, 428, 433, 434, 435, 437 (t), 439, 442, 446; Stephen Hutchings: pp, 4 (b), 13, 15, 18, 46, 80, 125, 158, 222, 247, 253, 271(t), 279, 291, 305, 312, 319 (t), 327 (t), 330 (t), 338, 341 (b), 344, 348, 353, 362, 363, 368 (t), 372, 382 (t), 385, 399, 402, 406, 408 (t), 421 (t); Jock MacRae: pp. 2, 26, 32, 61, 73, 74, 79, 83, 85, 111, 137, 296; Liz Milkau: pp. 81, 82, 330 (b), 403 (b); Dorothy Siemens: pp. 9, 90, 118, 171, 200, 201, 226, 235 (b), 316, 324, 325, 327 (b), 331, 340, 341 (t), 346, 356 (b), 358, 359, 368 (m), 369, 370, 375 (t), 379, 401, 408 (b), 409, 411, 412, 436, 437 (b), 438; Henry Van Der Linde: pp. 87; David Wysotski (Allure Illustrations): pp. 16, 27, 364.

Technical Illustrations: Jock MacRae
Icon Illustrations: Carl Wiens

The authors and publisher gratefully acknowledge the contributions of the following educators in the development of *Math Everywhere*:

Michael Beetham
TEACHER, Westmount Public School
Waterloo District School Board
Kitchener, Ontario

June Buick
VICE PRINCIPAL, Our Lady of Fatima School
York Catholic District School Board
Woodbridge, Ontario

Rita Cardarelli
ELEMENTARY CURRICULUM RESOURCE TEACHER
Ottawa-Carleton District School Board
Nepean, Ontario

Josephine Carnevale
TEACHER LIBRARIAN, St. Edith Stein School
Dufferin-Peel Catholic District School Board
Mississauga, Ontario

Joseph DiFrancesco
PRINCIPAL, Our Lady of Fatima
Brant Haldimand Norfolk Catholic District School Board
Brantford, Ontario

Dana Free
TEACHER, H. W. Knight Public School
Durham District School Board
Cannington, Ontario

Wendy Gallant
VICE PRINCIPAL/MATH SUBJECT LEADER
Algonquin and Lakeshore Catholic District School Board
Kingston, Ontario

Wes Hahn
SPECIAL ASSIGNMENT TEACHER
Hamilton-Wentworth District School Board
Hamilton, Ontario

Colleen MacDonald
CONSULTANT
Ottawa-Carleton Catholic District School Board
Nepean, Ontario

Judy Mendicino
TEACHER, Manchester Public School
Waterloo District School Board
Cambridge, Ontario

Debbie Schwantz
TEACHER, Woodland Park Public School
Waterloo District School Board
Cambridge, Ontario

Theresa Spencer
SPECIAL ASSIGNMENT TEACHER
Sudbury Catholic District School Board
Sudbury, Ontario

Dianne Phillips
CONSULTANT
Upper Canada District School Board
Prescott, Ontario

Mary Beth Yahn
TEACHER, Algonquin Ridge Elementary School
Simcoe District School Board
Midhurst, Ontario

Stephanie Bishop
TEACHER, Geary Elementary School
School District 17
Oromocto, New Brunswick

Susan Brims
TEACHER, West Dalhousie Elementary School
Calgary Public Schools
Calgary, Alberta

Bonnie Chappell
DIRECTOR, CURRICULUM AND INSTRUCTION
School District 57
Prince George, British Columbia

Cindy Coffin
MATH & LANGUAGE ARTS CONSULTANT
Saskatoon Catholic Schools
Saskatoon, Saskatchewan

Ruth LeBlanc
MATH MENTOR
Moncton School District 2
Riverview, New Brunswick

Denise McWilliams
TEACHER/CONSULTANT
River East School Division
Winnipeg, Manitoba

Darren McMillan
CONSULTANT
School District 43
Coquitlam, British Columbia

Suzanne Prefontaine
SPECIALIST MATH TEACHER,
Holyrood Elementary
Edmonton Public Schools
Edmonton, Alberta

John Price
CURRICULUM CONSULTANT
School District 39
Vancouver, British Columbia

John Pusic
COORDINATOR—INSTRUCTIONAL SERVICES
School District 35
Langley, British Columbia

Deb Scott
TEACHER, Hastings School
Louis Riel School District
Winnipeg, Manitoba

Joanne Stubbs
ELEMENTARY MATH AND SCIENCE CONSULTANT
Prince Edward Island Department of Education
Charlottetown, Prince Edward Island

Marilyn Wolstenholme
TEACHER, Florence Elementary School
Cape Breton-Victoria School Board
Florence, Nova Scotia

Tammy Wu
TEACHER, Caulfield School
School District 45
West Vancouver, British Columbia